HAZARDOUS MATERIALS DICTIONARY

SECOND EDITION

Hazardous Materials Dictionary

RONNY J. COLEMAN

TECHNOMIC
PUBLISHING CO., INC.

Lancaster · Basel

Hazardous Materials Dictionary (Second Edition)

a TECHNOMIC®publication

Published in the Western Hemisphere by
Technomic Publishing Company, Inc.
851 New Holland Avenue, Box 3535
Lancaster, Pennsylvania 17604 U.S.A.

Distributed in the Rest of the World by
Technomic Publishing AG
Missionsstrasse 44
CH-4055 Basel, Switzerland

Printed in the United States of America
10 9 8 7 6 5 4 3 2 1

Main entry under title:
  Hazardous Materials Dictionary (Second Edition)

A Technomic Publishing Company book
Bibliography: p.

Library of Congress Catalog Card No. 94-60735
ISBN No. 1-56676-160-3

This book is dedicated to the memory of Fire Chief Warren E. Isman, Fairfax County Fire Department. He was a person that recognized the need for training hazardous materials response team personnel long before it was fashionable, someone who pressed onward at every opportunity to assure that the issue was not obscured by political rhetoric, and who was lost to the fire service before he was through with the task.

INTRODUCTION

Over the last few decades the emphasis on the proper handling of hazardous materials has grown significantly. Loss of life and property used to be the primary reason that government and the private sector focused attention on improved rules and regulations to deal with a specific problem. Hazardous materials emergencies have resulted in the loss of lives of both emergency service and civilian personnel. However, the motivation for increased emphasis on hazardous materials more recently has included the priority for protecting the environment.

Proper response by emergency service personnel is absolutely essential to meeting all three goals. And, these personnel have been attempting to train and prepare for the next event as a top priority for almost all first responder organizations. Fire Chief Warren Isman, a leading authority on hazardous materials, once characterized the problem of keeping current in the field of hazardous materials as "a constant challenge of dealing with information overload and obsolescence at the same time."

This book is our second compilation of terms and phrases that have accumulated in the field of hazardous material handling. We have continued with a special emphasis on the terminology that relates to the communications needs of those who are responding to emergencies in these materials, such as fire protection, law enforcement, and emergency response teams. To a lesser degree we have provided some of the terms used in code enforcement and programs such as disclosure and leaking underground tanks.

The book is a compilation from a wide variety of sources. Some of the terms come from private industries that store, transport, and transfer hazardous materials, such as the trucking and railroad industries. Others come from the governmental and institutional organizations that provide insight, guidance, or have mandatory duties to perform. Examples of these are DOT, SARA, and state and local programs. Some are taken from incident management techniques that are used in managing emergencies involving these products.

As most everyone recognizes, language is a living phenomenon. Words take on new meanings. Synonyms are created to make images. New words are coined to describe new needs to communicate. Therefore, this glossary is not intended to be the final word on the terms. We have tried to keep track of new phrases, acronyms, and "buzzwords" as they have been created. But, realistically no one can really keep up. We hope that this book will be of assistance to those who have to work through the maze of complicated information exchange that makes up the field of hazardous materials management.

Finally, I would like to recognize the assistance of Chief Isman, who passed away last year, for his friendship and support in seeking out information to include in this book. I would also like to express my appreciation to Linda Gilman Hunter, who typed the input as it was collected; Battalion Chief John Eversole of the Chicago Fire Department, for his role in keeping the network alive; Fire Protection Specialist Julie Kunze of the Fullerton Fire Department, for advising me of legislative and regulatory changes in terminology; and all of the men and women of the emergency services and hazardous materials management field who are part of this emerging field of technology.

AAL *see* APPLIED ACTION LEVEL.

"A" End of Car (RR) The end opposite that on which the hand brake is mounted.

AA Aluminum Association.

AAS Atomic Absorption Spectrometry.

AAMI Association for the Advancement of Medical Instrumentation.

"A" Unit (RR) A diesel unit equipped with a cab and operating controls.

Abandon To cease efforts.

Abandoned Site An inactive hazardous waste disposal or storage facility that cannot be easily traced to a specific owner, or whose owner has gone bankrupt and subsequently cannot afford the cost of cleanup; a location where illegal dumping has taken place.

Abatement A method of reducing the degree or intensity of pollution such as the restoration, reclamation, or recovery of natural resources adversely affected by that pollution; also, the use of such a method.

AB Brake (RR) The current standard freight car air brake system; *see* AUTOMATIC AIR BRAKES.

AB Control Valve (RR) The operating valve of the AB freight car air brake. Controls the charging, application and release of the brakes.

Absolute Block (RR) A block that a train is not permitted to enter while it is occupied by another train.

Absolute Filter A filter having a removal efficiency of at least 99.97% for .03 micron particles. Often called a high efficiency particulate air filter or HEPA Filter.

Absolute Pressure Gauge pressure plus atmospheric pressure.

Absorb To soak up or drink in, as when a hazardous material is taken into a plant, textile, or soil.

Absorbed Dose The quantity of a material absorbed per unit of mass of tissue; unit of dose varies from chemical to chemical.

Absorbent Materials Any material used to soak up chemical materials; examples: sand, sawdust, commercial bagged clay, kitty litter, and zorbal.

Absorption   (1) Movement of a chemical into a plant, animal, or soil; compare to ADSORPTION. (2) Any process by which one substance penetrates the interior of another substance. In oil spill cleanup, this process applies to the uptake of oil by capillaries within certain sorbent materials; see CAPILLARY ACTION.

Absorptive Clay   A special type of clay powder that can soak up chemicals and hold them; often used to clean up chemical spills.

Acaricide   A pesticide used to control spiders, ticks and mites; miticide.

Accelerant   A chemical substance used to initiate or promote a fire. Flammable liquids are the most common types of accelerants.

Accelerator   A device utilized in the field of nuclear engineering to increase the velocity and energy of charged elementary particles, e.g., electrons or protons, through the application of electrical or magnetic forces. Types of accelerators include betatrons, Cockcroft-Walton, cyclotrons, linear accelerators, synchrocyclotrons, synchrotrons, and Van De Graaff generators.

Acceptable Risk   A risk judged to be outweighed by corresponding benefits, or one that is of such a degree that it is considered to pose diminished potential for adverse effects.

Access Control Point   The point of entry and exit that regulates access to and from control zones.

Access Road   Any passage providing access to a treatment, storage, or disposal area within a hazardous waste management (HWM) facility, suitable for use by transport vehicles and emergency vehicles in all types of weather.

Accident   An uncontrolled event which has the potential for damaging life or property; see INCIDENT.

Accident Mechanism   The series of events that culminate in the release of hazardous chemicals outside of their normal containment.

Accidental Explosion   Unintentional detonation/ignition of explosive or suspected explosive materials not associated with criminal activity; generally relates to some type of industrial or commercial activity.

Accumulate   To build up, add to, store, pile up.

Accumulative Pesticides   Pesticides that tend to build up in the tissue of animals or remain persistent in the environment.

Accuracy   Degree of agreement between a measured value and a true or expected value.

Acetylcholinesterase (AChE)   In pesticides, an enzyme that will most rapidly hydrolyze acetylcholine as substrate, will not hydrolyze most non-choline esters, is inhibited by excess substrate, and is derived primarily from nervous tissue.

ACGIH   see AMERICAN CONFERENCE OF GOVERNMENTAL INDUSTRIAL HYGIENISTS.

Acid   A hydrogen-containing compound that reacts with water to produce hydrogen ions; a proton donor; a liquid compound with a pH less than or equal to 2. Acidic chemicals are corrosive.

Acid Suits   Specialized protective clothing that prevents toxic or corrosive substances from coming into contact with the body of the wearer.

Acidify   Add acid to lower the pH of a substance.

Acre   43,560 square feet. Commonly used term when assessing involved areas of spills and leaks; equal in area to a plot 440 feet long by 99 feet wide, 209 feet by 209 feet, or 436 feet by 100 feet.

ACS   American Chemical Society.

Action   What an actor does or did during a particular event.

Action Level (AL)   A recommended acceptable limit for contamination of drinking water; similar to a tolerance level (TL), but not enforceable.

Activated Carbon   A highly absorbent form of carbon, used to remove odors and toxic substances from gaseous emissions or liquid effluents.

Activator   A material that is added to a pesticide to increase its toxicity.

Active Ingredients   A term used in the development of pesticides. Refers to the chemical that has the toxic potential. Active ingredients are listed, in order, on a pesticide label as a percentage of weight or pounds per gallon of concentrate. To be contrasted with inert ingredients.

Active Portion   A portion of a facility where hazardous waste treatment, storage, or disposal operations are being conducted subsequent to November 19, 1980 and which has not yet been closed.

Active Resources (ICS)   Resources logged and assigned to work tasks during an emergency.

Activity   (1) A set of actions directed towards an intended outcome or result. (2) Rate of decay of a radioactive sample; measured in curies, i.e., number of nuclear transformations per second.

Act of God   An unanticipated grave natural disaster or other natural phenomenon of an exceptional, inevitable, and irresistible character, the effects of which could not have been prevented or avoided by the exercise of due care or foresight.

Actor   A person who is involved in an activity and does something to effect an event.

Actual Dosage   The amount of a pesticide's active ingredient (not of a formulated product) that is applied to a specific area or target.

Acute   Of short or intense course, not chronic; of recent or sudden onset demanding urgent attention.

Acute Dermal Poisoning   A single dose of toxic chemicals absorbed through the skin in an amount capable of causing death.

Acute Effect   An adverse effect on a human or animal, generally after a single significant exposure, with severe symptoms developing rapidly and coming quickly to a crisis; see CHRONIC HEALTH EFFECT.

Acute Exposure   Exposure that is severe but short in duration.

Acute Health Effects   Health effects that occur or develop rapidly after exposure to a substance.

Acute Inhalation Poisoning   A single dose of toxic chemicals absorbed into the lungs in an amount capable of causing death.

Acutely Hazardous Waste   A waste considered to present a substantial hazard whether improperly managed or not. EPA includes in this category explosives, waste shown to be fatal to humans in low doses, and those shown in mammalian studies to have specific toxicities.

Acute Oral Poisoning   A single dose of toxic chemicals absorbed into the digestive track in an amount capable of causing death.

Acute Poisoning   A single exposure of a toxic substance of high toxicity that, if untreated, would be lethal.

Acute Radiation Exposure   Exposure to high radiation levels over a short period of time, usually 24 hours or less.

Acute Radiation Syndrome   Illness that develops as a result of acute radiation exposure; consists of cerebrovascular, gastrointestinal, and hematopoietic syndromes.

Acute Release   Release of a hazardous material that is severe but short in duration.

Acute Toxicity   Harmful effects produced by a single short-term exposure that can result in biological harm or death; toxic symptoms that develop shortly after exposure, usually within 24 hours.

Additive   Any substance added to a pesticide to improve performance; same as ADJUVANT.

Additive Effects   Mixture of insecticides in which the total toxicity equals the sum of the individual insecticides.

Adenosine Triphosphate (ATP)   A compound present in living cells that provides energy derived from food or sunlight for the energy-expending process (such as protein synthesis, muscle contraction, impulse conduction, and glandular secretion).

Adherence   The characteristic of a material causing it to stick to another surface; a factor in contamination.

Adhesive   An additive or adjuvant that helps a pesticide adhere to different surfaces.

Adhesion   The attraction of unlike molecules. Molecular attraction that holds the surfaces of two substances in contact, such as water and rock particles.

Adiabatic Ignition   A rapid compression of flammable vapors that generates a sufficient amount of heat to cause the ignition of those vapors. Synonymous with dieseling.

Adjuncts (EMS)   Equipment, special devices, and drugs used by specially trained personnel to assist in performance of life support measures. Examples: airways, cardiac monitors, intravenous infusion, oxygen, lidocaine.

Adjustable Undercarriage   A truck undercarriage with a provision for convenient fore-and-aft adjustment of its location on the trailer; used to shift a greater part of the vehicles' gross weight or load onto the kingpin or suspension.

Adjuvant   Any substance (usually an inert ingredient) added to a pesticide solution to make the formulation work better. Synonyms: adhesive, emulsifier, penetrant, spreader, wetting agent; same as additive.

Administering Agency   The designated unit of a county or city tasked with administering the local implementation of the state and federal hazardous material emergency planning and community–right-to-know programs.

Adsorb   To collect, as a gas or liquid, in condensed form on the surface of another substance.

Adsorption   (1) The process by which one substance is attracted to and adheres to the surface of another substance without actually penetrating its internal structure. (2) Process by which a substance is held (bound) to the surface of a soil particle or mineral in such a

5

way that the substance is only available slowly. Clay and highly organic materials tend to adsorb pesticides rather than absorb them.

Adulterated   Contaminated or made impure.

Advance of a Signal (RR)   The side of the signal opposite to that from which the indication is received.

Advanced Life Support (ALS)   Emergency medical treatment at advanced levels which includes the administration of drugs, intravenous fluids, cardiac monitoring/intervention, and other medical care provided by paramedics in the prehospital setting.

AENOR   Asociacion Espanola de Normilizacion y Certificacion (Spain).

Aerator   A device used in the moving of dry bulk solids in order to improve the flowability of the material.

Aerobic   Living in the air; opposite of anaerobic.

Aerosol   A system in which liquid or solid particles are distributed in a finely divided state through a gas, usually air. Particles within aerosols are usually less than 1 micron (0.001 mm) in diameter, and are more uniformly distributed than in a spray.

Aerosol (Generator)   A low concentration solution of a pesticide or combination of pesticides, usually in the form of oil solution especially formulated for use in generators.

Aerosol (Pressurized Can)   A small amount of material of any sort driven through a fine opening by an inactive gas under pressure. When nozzle is triggered it produces a fine spray, mist, or fog of minute solid or liquid particles suspended in the air.

AFFF   Aqueous Film Forming Foam; an extinguishing agent that may be used on many flammable liquids.

AFNOR   Association Française de Normalisation (France).

After Action Report   A post-incident analysis report generated by a responsible party or responding agency after termination of a hazardous material incident; describes actions taken, materials involved, impacts, etc.

AGA   American Gas Association.

Agency Specific Plan   An emergency plan written by and addressing an individual agency's response actions, capabilities, and resources.

Agent (RR)   Freight agent.

Agitate   To keep a chemical formulation mixed up; to keep a formulation from separating or settling out in a tank.

Agitation The act of mixing or stirring a chemical formulation.

Agitator The paddle or other mechanical device that uses air, hydraulic action or some other form of movement to keep a chemical formulation mixed in a vessel.

Agricultural Commodity Any plant, animal, or part of a plant, animal, or animal product that is to be bought or sold in commerce.

Agricultural Waste Waste materials such as rice straw, wood wastes, orchard prunings, bark, manure, and cotton gin trash.

Agroecosystem The system of plants, animals, and habitats used to serve human purposes.

Agronomist Scientist specializing in study of soil or plants as related to crop or crop production.

AIA Aerospace Industries Association (American Insurance Association).

AIDS Acquired Immune Deficiency Syndrome; a communicable disease caused by Human Immunodeficiency Virus (HIV), a biohazard.

AIDS-Related Complex (ARC) A term to describe symptoms of HIV infection in people who have not yet developed AIDS. "HIV positive" has generally replaced "ARC."

AIHA *see* AMERICAN INDUSTRIAL HYGIENE ASSOCIATION.

Air Bill Shipping papers prepared from bill of lading when hazardous materials are moved by an air carrier.

Air Blast Sprayer Machine used in pesticide applications to orchards, shade trees, vegetables, and for fly control.

Airborne Any substance transported by wind or wind currents.

Airborne Pathogen Pathologic microorganisms spread by droplets expelled into the air, typically through a productive cough or sneeze.

Airborne Radioactives Any radioactive material dispersed in the atmosphere in the form of dusts, fumes, vapors, or gases.

Air Brake Hose (RR) Flexible connection between brake pipes of railroad cars or locomotives.

Air Compressor (RR) Power-driven air pump supplying compressed air for operation of air brakes and other air-actuated equipment.

Air Connection Fitting used to take air pressure from source to another pressure vessel; used on trucks, tanks, and other cargo containers.

Air Gage (Gauge) Instrument indicating amount of air pressure in reservoirs or brake pipes.

Air Inversion Meteorological condition in earth's atmosphere in which the air some distance from earth surface is higher in temperature than that at ground level. Such condition traps air and released vapors near the earth surface, thereby impeding dispersion.

Air Lift Axle A single air-operated axle that will convert vehicle into multi-axle unit when lowered, providing vehicle with much greater load capacity. Trucking industry term.

Air Modeling Mathematical models used to predict movement and concentrations of chemicals in the atmosphere.

Air Monitoring To observe, record, and/or detect pollutants in ambient air.

Air Pipe (Air Brake) *see* BRAKE PIPE.

Air Pollution Contamination of atmosphere with any material that can cause damage to life or property.

Air Pollution Control Districts/Air Quality Management Districts (APCDs/AQMDs) County and regional agencies established pursuant to the County Air Pollution Control Law of 1947 and the Mulford-Carrell Act of 1967 to administer and enforce minimum standards for air quality in the state of California.

Air Purifying Respirators (APR) A breathing mask with specific chemical cartridges designed to either filter particulates or absorb contaminants before they enter the worker's breathing zone. Intended to be used only in atmospheres where the specific chemical hazards are known; *see* PERSONAL PROTECTIVE EQUIPMENT. For higher level of protection; *see* SELF-CONTAINED BREATHING APPARATUS.

Air Purifying Respirator — Powered An APR with a portable motor to force more air through the filtering/purifying cartridges; can only be utilized in atmospheres where the chemical hazards are known.

Air Quality Management District (AQMD) A local/regional air pollution agency responsible for regulation and monitoring of air quality: *see* AIR POLLUTION CONTROL DISTRICTS/AIR QUALITY MANAGEMENT DISTRICTS.

Air-Reactive (Materials) Substances that will ignite into open flame when exposed to open air.

Air Sampling Collection and analysis by instrument of samples of air to determine presence of hazardous materials.

Air Spring (Air Bags) Flexible air-inflated chamber which controls air pressure and can be varied to support load and absorb road shocks. Can fail without warning, creating unstable load.

Air (or Water) Streams Method of oil containment where force of air or water directed as a stream can be used to divert or contain an oil slick.

Air Support Supervisor (ICS) Personnel responsible for development and implementation of air requirements of an emergency; reports directly to operations section chief.

Air Tanker (ICS) Any fixed-wing aircraft certified by FAA as being capable of transport and delivery of fire retardant solutions.

AL Action level.

AL A suffix indicating the presence of an alcohol or aldehyde group, as in choral or pentenal.

Alcohols A class of organic chemical compounds containing hydroxyl group, used as solvents and in some preparations of chemical dispersants; see GLYCOLS.

Alertor (Deadman Control) (RR) Device that detects frequency of engineman's movements and initiates an air brake application when required frequency is not maintained; see DEADMAN CONTROL.

Algicide A pesticide used to control algae, especially in stored water or industrial applications.

Aliphatic Carbon atoms linked in a chain-like formation; includes alkanes and alkenes.

Aliquot Dividing a sample into two or more equal parts; implies an exact division of a quantity. An aliquot of a field sample (soil or water) is often used for analysis in a laboratory.

Alkali Any compound which forms the hydroxyl ion in its water solution; see BASE. Synonyms include hydroxide, caustic.

Alkalinity The concentration of hydroxide ions.

Alkaloids Chemicals present in some plants, sometimes used as pesticide.

Alkanes A class of hydrocarbons that can be gas, solid, or liquids depending upon carbon content; its solids (paraffins) are a major constituent of natural gas and petroleum. Alkanes are usually gases at room temperature (methane) when containing less than five carbon atoms per molecule. Low carbon number alkanes produce anaesthesia and narcosis at low concentrations and at high concentrations can cause cell damage and death in a variety of organisms. Higher carbon number alkanes are not generally toxic but have been shown to interfere with normal metabolic processes.

Alkenes A class of hydrocarbons; also called olefins; sometimes are gases at room temperature but are usually liquids; common in petroleum products. Generally more toxic than alkanes, but less toxic than aromatics.

Alley (slang) (RR) A clear RR track for movement through a yard.

Allo- A prefix indicating the compound to be a close relative, an isomer or a variety of the compound to the name of which it is affixed, as in allocaffeine. For example, ethylene somerist is prefixed to the name of the more stable form into which the compound can be converted on heating; thus, fumaric acid is allomaleic acid since it is the stable isomer on heating maleic acid.

Alloy Mixture of two or more metals.

Alpha Particle A positive-charged particle emitted by certain radioactive materials. Consisting of two neutrons and two protons, is identical with nucleus of the helium atom. Least penetrating of the three common forms of radioactive substances (alpha, beta, gamma), it is not normally considered dangerous to plants, animals, or people unless it gets into the body.

Alternative Technology The application of technology to the reduction of waste generation, promotion of recycling, and alternatives to land disposal of hazardous waste.

Alum Usually potassium aluminum sulfate used in water-treatment plants for settling out small particles of foreign matter; usually gelatinous when wet.

AMA Automobile Manufacturers Association (American Medical Association).

Ambient Surrounding conditions, primarily used in reference to climatic conditions, such as ambient temperature.

Ambient Air Quality Quality of the surrounding atmosphere or circulating air.

Ambient Air Quality Standards Specified concentrations and durations of air pollutants reflecting relationship between the intensity and composition of air pollution to undesirable effects as established by a state board and/or the the federal government.

Ambulance Service Area (EMS) Designated geographic area delineated by the local EMS agency to ensure availability of emergency medical transport services at all times by one or more specified providers.

American Conference of Governmental Industrial Hygienists (ACGIH) Professional society of persons employed by official governmental units who are responsible for full-time programs of industrial hygiene. Primary function is to encourage the interchange

10

of experience among governmental industrial hygienists and to collect and make available information of value to them. Also promotes standards and techniques in industrial hygiene and coordinates governmental activities with community agencies.

American Industrial Hygiene Association (AIHA)   Organization of professionals trained in the recognition and control of health hazards and the prevention of illness related thereto. Promotes the study and control of environmental factors affecting the health of industrial workers; provides information and communication services pertaining to industrial hygiene.

American National Standards Institute (ANSI)   Industrial firms, trade associations, technical societies, consumer organizations, and government agencies cooperating to create nationally coordinated voluntary safety, engineering, and industrial standards; serves as a clearing house.

American Society for Testing and Materials (ASTM)   Establishes voluntary consensus standards for materials, products, systems, and services. Sponsors research projects, develops standard test methods, specifications, and recommended practices now in use.

Ammonia Stripping   A steam process used in treatment of ammonia-bearing wastes. Process also used to remove volatile and organic contaminants from the waste stream. Ammonia readily condenses and can be reclaimed for use.

Anaerobic   Able to live where there is no oxygen; opposite of AEROBIC.

Analeptics   Medullary-stimulating drugs which increase respiration and blood pressure in mammals.

Analog   A chemical compound similar in structure to another chemical compound but differing in some slight structural detail.

Analysis   The separation of a compound into its constituent parts.

-ane   A suffix usually indicating a saturated (paraffin) hydrocarbon, as in pentane. Also used to indicate saturated parent compounds in the cyclic series, as in camphane or dioxane.

Angle Cock (RR)   A two-position valve located at both ends of the air brake on railroad cars. When open, allows passage of air.

Anhydrous   Without water, dry.

Aniline Point   The lowest temperature at which a chemical aniline and a solvent (such as the oil in oil-based muds) will mix completely.

Annular Seal   Impermeable material, such as cement grout, that fills the space between a borehole and a blank well casing.

11

ANS   American Nuclear Society.

ANSI   see AMERICAN NATIONAL STANDARD INSTITUTE.

Antagonism   Decrease in effectiveness resulting from two or more chemicals being mixed together.

Anti-   A prefix used in certain cases to indicate an opposed arrangement of groups of atoms, as in antiartaric acid.

Antibiotic   Substance used to control pest microorganisms.

Antibody   A component of the immune system that eliminates or counteracts a foreign substance (antigen) in the body.

Anticoagulant   Chemical (usually a pesticide) preventing the normal clotting of blood.

Antidote   Immediate treatment given to counteract effects of poisoning.

Antigen   A foreign substance that stimulates the production of antibodies in the immune system.

Antitranspirant   A pesticide that reduces water loss by coating the leaves of a plant.

AOSM   Arab Organization for Standardization and Metrology.

Aphicide   Pesticide used to control aphids.

API   American Petroleum Institute.

API Gravity   A scale developed by the API designating an oil's specific gravity or the ratio of weight of oil to pure water.

Application   Directing a pesticide onto plants, animals, buildings, soils, water, or any other area.

Applicator   A person or piece of equipment that applies pesticides.

Applied Action Level (AAL)   A criterion that delineates a concentration of a substance (e.g., benzene) in a medium (e.g., water) which, when exceeded, is determined to present a significant risk of an adverse impact to a biological receptor.

Approach Signal (RR)   Fixed signal preceding an interlocking RR signal and governing the approach to the interlocking.

Apron Ring   The lowest ring of plates in an oil tank.

Apo-   A prefix indicating a compound which is related to, or formed from, the compound to the name of which it is prefixed, as in apomorphine.

AQMD *see* AIR QUALITY MANAGEMENT DISTRICT

Aqueous Indicates water is present in the solution.

Aqueous Treatment A hazardous waste treatment system designed to remove contamination from water so it may be safely returned to the environment.

Aquifer Underground zone of earth containing water; frequently is source of water supply (e.g., wells); usually composed of gravel or porous material and can be contaminated by waste disposal or chemical releases.

Aquitard (Aquiclude) An underground geological formation that is impermeable to or impedes the movement of water.

ARB Air Resources Board (state).

Arbitrary and Capricious Without fair, solid, and substantial reason.

Area Plan A plan established for emergency response to a release or threatened release of a hazardous material within a city or county. (California Health and Safety Code, Section 25503, Chapter 6.95)

Arenaceous Pertaining to sand or sandy rocks, such as shale.

Argillaceous A non-productive formation consisting of clay or shale.

ACRI Air Conditioning and Refrigeration Institute.

Aromatics Class of hydrocarbons considered to be the most immediately toxic; found in oil and petroleum products; soluble in water; considered long-term poisons and carcinogens.

AROS African Regional Organization for Standardization.

Arrival Notice (RR) Notice furnished to consignee regarding freight.

ARS Agricultural Research Service; a division of USDA.

Arsenicals Pesticides containing arsenic.

Arson Deliberate damage to or destruction of property through use of an incendiary device, destructive device, or explosive that falls within the purview of ATF.

AS Asymmetrical; abbreviation used to indicate an asymmetrical arrangement of substitution of a parent compound.

ASA Acoustical Society of America.

ASAE American Society of Agricultural Engineers.

13

Asbestos   Any material containing more than one percent asbestos in any form.

Asbestosis   A disease of the lungs caused by the inhalation of fine airborne fibers of asbestos.

ASCE   American Society of Civil Engineers.

ASChE   American Society of Chemical Engineers.

-ase   A suffic used in forming the name of an enzyme. Usually the suffix is added to the name, or a part of the name, of a substance upon which the enzyme acts, such as maltase.

Aseptic   Free of disease-causing organisms.

Ash   The incombustible material remaining after a fuel or solid waste has been burned.

ASME   American Society of Mechanical Engineers.

ASME Container   Container constructed in compliance with ASME codes.

ASMW   Amt für Standardisierung, Messewsen (Germany).

Asphalt   Hydrocarbon material ranging in consistency from heavy liquid to a solid. Most common source is residue left after fractional distillation of crude oils; used primarily for surfacing roads.

Asphyxiant   A vapor or gas that can cause unconsciousness or death by suffocation (lack of oxygen).

Asphyxiating Materials   Substances causing death by displacing of oxygen in the air; example: carbon dioxide.

ASQC   American Society for Quality Control.

Assessment   The process used to determine the nature and degree of a hazardous material or a hazardous materials incident from a safe vantage point by any means possible.

Assigned Car (RR)   A car that has been filled with a commodity and given a specific destination on the waybill or consists.

Assisting Agency (ICS)   Any agency directly contributing suppression, rescue, support, or service resources to another authority having jurisdiction.

Associated Gas   Natural gas, commonly known as gas-cap gas, which overlies and is in contact with crude oil in the reservoir.

Association of American Pesticide Control Officials, Inc.   Consists of officials charged by law with active execution of the laws regulating the sale of economic poisons, and of deputies designated by these officials employed by state, territorial, dominion, or federal agencies.

Association of American Railroads (AAR)   A central coordinating and research agency of the American railway industry.

ASTM   *see* AMERICAN SOCIETY FOR TESTING AND MATERIALS.

Asymptomatic   Without symptoms.

ATA   American Trucking Association.

-ate   A suffic used to denote an ester or a salt, such as in ethyl acetate or sodium acetate.

ATF   Alcohol, Tobacco, and Firearms division of the United States Department of the Treasury.

Atom   A basic unit of physical matter indivisible by chemical means; the fundamental building block of chemical elements; composed of a nucleus of protons and neutrons, surrounded by electrons.

Atomic Cloud   The hot gases, smoke, dust, and other matter carried aloft after an explosion of a nuclear weapon in the air or near the surface; frequently has a mushroom shape.

Atomic Energy Commission (AEC)   The independent civilian agency of the federal government with statutory responsibility for atomic energy matters; the body of five persons appointed by the president to direct the agency.

Atomic Number   Number of protons in the nucleus of an atom. Each chemical element has been assigned an atomic number in a complete series from 1 (hydrogen) to 103 (lawrencium), corresponding to the number of protons in its nucleus.

Atomic Weapon   An explosive weapon packaged for transportation or use by military forces in which the energy is produced by nuclear fission or fusion, as opposed to a nuclear device used for peaceful purposes such as energy production.

Atomic Weight   The mass of an element relative to its atoms.

Atomize   To break up a liquid into very fine droplets by forcing it through a nozzle-like device.

Atropine   Antidote used by medical profession to treat poisoning by organic phosphates or carbamates.

Attack Line   Hose lines used to apply water or foam to a fire, generally 1-1/2 inches or 2-1/2 inches in diameter and flowing between 60 and 250 gpm.

Attapulgite   A highly sorptive natural clay mineral having fibrous or spicular particles.

15

Attempted Bombing (Failure to Detonate)   Unsuccessful attempt to commit a bombing with high or low explosives or blasting agents, due to the malfunction, recovery or disarmament of an explosive device.

Attempted Incendiary Bombing (Failure to Ignite)   An unsuccessful attempt to commit an incendiary bombing due to the malfunction, recovery, or disarmament of the device.

Attractant   A chemical that lures insects, rodents or other pests to selected locations where they can be destroyed.

Autoecology   The branch of ecology dealing with the interrelationships of the individual organism and its environment.

Authority (EMS)   Governmental or legal body or a representative of that organizations having legal responsibility at a particular incident; established by Division 2.5, Section 1797.100 of the Health and Safety Code.

Authority Having Jurisdiction (AHJ)   The organization, office, agency, entity, corporation, or person responsible for approving an action within a specific jurisdictional boundary. The AHJ is responsible for providing for the position of incident commander/scene manager at the scene of a hazardous materials incident that occurs within their jurisdictional response boundaries.

Authority to Construct   Permit issued prior to construction of facilities that will emit a significant amount of pollutants to the atmosphere.

Authorized Registered Nurse (EMS)   Personnel certified as qualified to provide emergency care and to issue emergency medical instructions to EMTs.

Authorized Representative   Individual responsible for overall operation of a facility or a unit of that facility.

Auto-Ignition Temperature   Minimum temperature to which a substance must be heated to initiate self-sustained combustion, independent of any open flame.

Automatic Aid   A preplanned system where two or more fire departments respond routinely across jurisdictional boundaries to render mutual assistance; see MUTUAL AID AGREEMENT.

Automatic Air Brake (RR)   A fail-safe device that prevents a runaway train. It consists of an activation device that checks on the status of the train engineer. If the operator of the train is inactive for a specified period of time the brakes are automatically applied; see ALERTOR, DEADMAN CONTROL. Sometimes called ATS (Automatic Train Stop).

16

Automatic Control Device regulating various factors (such as flow rate, pressure, or temperature) of a system without supervision or operation by personnel; *see* SDV and PSH & L.

Autoradiograph A photographic record of radiation from radioactive material in an object, made by placing the object very close to a photographic film or emulsion; used to locate radioactive atoms or tracers in metallic or biological samples.

Available Resources (ICS) Personnel and material assigned to an incident, available for assignment.

Avicide A pesticide used to control birds.

Avoid Prolonged Contact Phrase commonly used on pesticide labels warning against inhalation of vapors.

AWS American Welding Society.

AWWA American Water Works Association.

Axle A beam with spindles about which the wheels of a vehicle rotate.

Axle Camber The controlled convexity of a trailer axle allowing for deflection of the axle under the load, compensating for the curvature in the crown of road beds.

Axle Setting The distance from the centerline of a truck axle to the trailer rear surface.

Axle Weight Amount of weight transmitted to the ground by one axle or the combined weight of two axles in a tandem assembly.

Axon A nerve cell process usually singular and elongate, terminating in short branches.

B

"B" End of the Car (RR) The end on which the hand brake is located.

"B" Unit (RR) A diesel unit without a cab and without complete operating controls. Usually equipped with hostler controls for independent operating at terminals.

BADCT Best Available Demonstrated Control Technology.

Back End Loader Refuse truck that has power-driven loading equipment at the rear of the vehicle.

Background Radiation The radiation in our natural environment, including cosmic rays and radiation from the naturally radioactive elements, both inside and outside the bodies of people and animals; also called natural radiation.

Back Haul (RR)   To haul a RR shipment back over part of the route which it has traveled.

Backpressure   That pressure resulting from restriction of full natural flow of oil or gas.

Backscatter   When radiation of any kind strikes matter (gas, liquid, or solid), some of it may be reflected or scattered back in the general direction of the source; important when beta particles are being counted.

Backshore   Area of the shoreline above the high tide mark, only inundated with water during exceptionally high tides accompanied by high winds; does not support characteristic flora and fauna. Can supply granular materials for replacement of oil-contaminated beach material during shoreline cleanup programs.

Bacteria   A type of living one-celled microorganism that can produce disease in a suitable host. Bacteria can self-reproduce, and some forms may produce toxins harmful to their host.

Bactericide   A pesticide used to control disease-causing organisms.

Backwash   Reversal (downward) of waterflow in well to remove fines and enhance production.

Bad Order (RR)   RR car in need of repair.

Baffle   An intermediate partial bulkhead in a tank truck which reduces the surge effect in a partially loaded truck.

Bag In–Bag Out   Method of removing items from a contaminated enclosure utilizing plastic bagging material to prevent spread of contamination.

Baler   Machine used to reduce volume of waste materials, usually by baling them into rectangular bundles, for shipping, recycling, or landfill.

Ballast (RR)   Material placed on a RR roadbed to hold track in line.

Ballast Car   A car used for carrying ballast material to repair site, usually a gondola or hopper.

Ballast Tamper   A machine that compacts ballast under RR ties.

Barrel   Unit of liquid measure used extensively by petroleum industry; equal to thirty-five Imperial gallons or 160 liters.

Barricade Shield   A type of movable protection from radiation.

Barrier (Containment Barrier) Any non-floating structure constructed to contain or divert spilled material.

Barrier Well A well installed to intercept and pump out a plume of contaminated ground water.

Base (1) Substance containing group forming hydroxide ions in water solution. (2) (ICS) A location where primary logistics functions are coordinated and administered during an emergency incident; an incident command post or command post.

Base Manager (ICS) Person responsible for activation of the incident command post (base), supervises base operations, distributes working materials for staff, and provides food, sleeping, and sanitation facilities for assigned personnel.

Base Hospital (EMS) One of a limited number of hospitals entering into a written contractual agreement with the local EMS agency responsible for directing the advanced life support system or limited advanced life support system assigned to it.

Base Soils Unconsolidated material (sand, silt, gravel, etc.) that separates the lower limits of refuse from groundwater and bedrock.

Basic Life Support Emergency care treatment at a level authorized to be performed by emergency medical technicians as defined by the authority having jurisdiction. Care is generally limited to stabilization of respiratory and circulatory functions and, if needed, the immobilization of the cervical spine. Does not include use of drugs or advanced telemetry equipment. Commonly called EMT-1 or EMT-A.

Basin Any uncovered device used to retain wastes as part of a treatment process, usually holds less than 100,000 gallons.

Basin Plan A water quality control plan adopted by a Regional Water Quality Control Board and approved by the State Water Resources Control Board that includes actual and potential uses of ground and surface water and water quality objectives to protect the designated uses (California term).

Beam A stream of particles or electromagnetic radiation going in a single direction.

Becquerel (BQ) The SI unit for measuring the activity of a radioactive sample.

Belt Line (RR) A short railroad operating within a city.

Bentonite Grout An aluminum silicate clay, to which a small amount of magnesium oxide is added, that swells and forms a viscous suspension when mixed with water. Upon drying, it forms a hard cement–like material. Commonly used to refill and seal soil coring holes, as a fill or grout material around well casings, or to fill and seal off abandoned wells.

Berm A ledge or shoulder, as along the edge of a paved road.

Beta A type of radiation, essentially an electron or positron, which can cause skin burns. Beta emitters are harmful if they enter the body but can be shielded by protective clothing.

Beta Particle An elementary particle emitted from a nucleus during radioactive decay.

bi- A prefix indicating the number two.

Bill of Lading Commercial document that accompanies a shipment of materials and lists all items in the shipment.

Binding Arbitration A process for the resolution of disputes wherein decisions are made by an impartial arbitrator; decisions of that arbitrator are binding.

Bioaccumulation Process occurring when toxic substances are passed through the food chain from soil to plants to grazing animals to human beings. Also the absorption and storage of toxic chemicals from the environment in the body, usually in body fat.

Bioassay The utilization of living organisms to determine the biological effect of some substance, factor, or condition. Determination of the relative strength and toxicity of a substance, such as a drug, by comparing its effect on a test organism with that of a standard preparation.

Biochemical Oxygen Demand (BOD) The amount of oxygen required by bacteria to stabilize decomposable organic matter under aerobic conditions.

Bioconcentration Process of a pesticide being concentrated in the tissues of plants or animals.

Biodegradable Waste material capable of being broken down by bacteria into basic elements.

Biodegradation The breaking down of substances into basic elements.

Biohazard Infectious agents presenting a risk or potential risk to living organisms, either directly through infection or indirectly through disruption of the environment.

Biohazard Area Any area in which work has been or is being performed with biohazardous agents or materials.

Biological Agent Microorganisms (primarily bacteria) added to the water column or soil to increase the rate of biodegradation of spilled oil. Also, nutrients added to water to increase growth.

Biological Control   The use of living organisms to control pests; opposite of using pesticides.

Biological Half-Life   The time required for a living organism to eliminate half of a substance which it takes in.

Biological Hazardous Wastes   Any substance of human or animal origin, other than food wastes, which is to be disposed of and could harbor or transmit pathogenic organisms; includes, but is not limited to, pathological specimens such as tissues, blood elements, excreta, secretions, bandages, and related substances.

Biological Magnification   The concentration of certain substances in a food chain; an important mechanism in the concentration of pesticides and heavy metals in organisms such as fish.

Biological Material   An organism or a group of organisms that make up a certain material.

Biological Productivity   The rate at which the energy of the sun is transferred to and reflected in growth and/or abundance of plants and animals; an important consideration in assigning priorities for hazardous materials spill cleanup.

Biological Transformation   Structural alteration of a chemical by an organism. In regard to fuel, refers primarily to the decomposition of organic compounds by microorganisms.

Biological Treatment   Process by which hazardous waste is rendered non-hazardous or reduced in volume by the actions of microorganisms. The five principal techniques include activated sludge, aerated lagoon, trickling filters, waste stabilization ponds, and anaerobic digestion.

Biomagnification   Concentration of a pesticide in tissues of plants or animals; same as bioconcentration.

Biomedical Telemetry (EMS)   Transmission of biological data from a patient to a monitoring point by means of radio or wired circuits.

Blasting Agents   Any material or mixture consisting of fuel and oxidizer intended for blasting, not otherwise defined as an explosive.

Blast Wave   A pulse of air propagated from an explosion in which the pressure increases sharply at the front of a moving air mass, accompanied by strong, transient winds; shock wave.

Bleed Valve   A mechanical valve used to release liquid or gas pressure in a system. Bleed valves are different from pressure relief valves in that bleed valves must be operated manually.

BLEVE   Boiling Liquid Expanding Vapor Explosion. Normally associated with fires that involve compressed gases in cylinders; rapid rupture of a vessel caused by over-pressure accompanied by rapid burning of the tank contents.

Blind Sample   A sample whose composition or source is known to the agency but not known by the person logging in samples or the analyst. It is submitted along with the regular field sample set. When both the anticipated sample composition and the blind status of the sample are not known to the analyst, the sample is called a "double blind" sample. A blind sample is used to check analytical performance and proficiency.

Bloodborne Pathogen   Pathologic mircoorganisms that are present in human blood that can cause disease in humans (OSHA). Note: the term "blood" includes blood, blood components, and products made from human blood.

Blow-Down Valve   Manually operated valve on a cargo tank that quickly reduces the tank pressure to the atmospheric pressure.

Board (RR)   (1) A fixed signal regulating railroad traffic and usually referred to as a slow board, clear board, order board, or red board. (2) A list of employees available for service, also called Extra Board.

Boarding Car (RR)   A car used as a place of lodging for workmen.

Body Center Plate   The center plate attached to the underside of a truck body bolster; *see* CENTER PLATE.

Body Fluids   Fluids that have been recognized by the Centers for Disease Control (CDC) as directly linked to the transmission of HIV and/or HBV and/or to which Universal Precautions apply: blood, semen, blood products, vaginal secretions, cerebrospinal fluid, synovial fluid, pericardial fluid, amniotic fluid, and concentrated HIV or HBV viruses (OSHA).

Body Substance Isolation (BSI)   An infection-control strategy that considers all body substances potentially infectious; *see* UNIVERSAL PRECAUTIONS.

Bogie (RR)   (1) The running gear of a highway semi-trailer which may be removed. (2) A swivel railway truck.

Boiling Point   The temperature at which a substance changes from a liquid to a gas.

Boiling Water Reactor   A nuclear reactor in which water, used as both a coolant and a moderator, is allowed to boil in the core; resulting steam is used to drive a turbine.

Bolster   Cross member on underside of truck body through which the weight is transmitted.

Boom (Containment Boom)   A floating mechanical structure that extends above and below the water surface, designed to stop or divert the movement of an oil slick.

Boom Failure   Failure of a containment boom due to improper deployment or excessive winds, waves, or currents.

Bootie   A sock-like over-boot protector worn to minimize contamination.

Borehole   An uncased well drill hole.

Boring   A hole created by a drilling device.

Botanicals   Pesticides made from plants; examples: nicotine, pyrethrum, rotenone, and strychnine.

Bottom Plug   The plug or cap placed at the bottom of a well casing to preclude erosion of the backfill material.

Box Car (RR)   An enclosed RR car used for general service, especially for cargo that must be protected from the weather.

Boyle's Law   A law in chemistry relating to the condition of gases, PV = Constant; volume of a given mass of gas varies inversely with the absolute pressure if temperature remains constant.

Brake Club (RR)   Stick used by freight trainmen to tighten hand brakes.

Brake Cylinder (Air brakes)   Cylinder containing a piston forced outward by compressed air to operate brakes. When air pressure is released, piston is returned to its normal position by a release spring coiled about the piston rod.

Brakeman   RR employee who assists with train and yard operations.

Brake Pawl (Hand brake)   A small, specially shaped steel piece pivoted to engage teeth of brake ratchet wheel to prevent turning backward, thus releasing brakes.

Brake Pipe   Air brake piping of a locomotive or RR car acting as a supply pipe for the reservoirs.

Brake Ratchet (Hand brake)   Wheel attached to the brake shaft with teeth which the pawl engages.

Brake Shaft   A shaft on which a chain is wound and by which the power of a hand brake is applied to the wheels.

Brake Shoe   Friction material shaped to fit the tread of the wheel when the brakes are applied.

Brake Valve (Air brake)   The valve by which the engineer operates the brakes.

Brake Wheel   An iron wheel attached to upper end of the brake shaft which is turned manually to apply the brakes.

Brand Name   Name given to a chemical product by the manufacturer; may or may not reflect the characteristics of the material.

Breakthrough Time   The elapsed time between initial contact of a hazardous chemical with the outside surface of a protective clothing material and the time at which the chemical can be detected at the inside surface of the material.

Breathing Apparatus   A self-contained, compressed air, pressure/demand (positive pressure) respiratory device designed to provide full respiratory protection; seals the airway against contamination; see SELF-CONTAINED BREATHING APPARATUS (SCBA).

Breathing Zone Air Sample   A sample collected in the breathing area of a worker to assess exposure to airborne contaminants.

Breeder Reactor   A nuclear reactor producing fissionable fuel as well as consuming it; especially any reactor that produces more fuel than it consumes.

Bremsstrahlung   Electromagnetic radiation emitted when a fast-moving charged particle loses energy upon being accelerated and deflected by the electric field surrounding a positively charged atomic nucleus. S-rays produced in an ordinary X-ray machine are bremsstrahlung.

Brine   Water saturated with salt.

Brine Channels   Small passages formed in the lower surface of first-year sea ice by the exclusion of saline water during rapid freezing.

BSI   British Standards Institution (United Kingdom).

BTU   British Thermal Units, a measuring unit of heat.

Brush Patrol Unit (ICS)   A light mobile unit having limited pumping water capacity, capable of off-road operations.

BTX   Benzene, toluene, and xylene; term used in gasoline analysis.

BTX&E   Benzene, toluene, xylene, and ethylbenzene.

Bubble Barrier   A barrier used with some success to contain oil-spills in calm harbors.

Bubble Chamber   A device used for detection and study of elementary particles and nuclear reactions.

Buddy System   A system of organizing employees into work groups in such a manner that each employee of the work group is designated to be observed by at least one other employee in the work group.

Buffer Zone   Minimum acceptable space between active portion of a hazardous waste facility and the property line. Usually under control of state or local standards.

Buildup   Accumulation of a hazardous chemical; usually applied to pesticides.

Bulk Container   A cargo container, such as may be transported by truck or railroad; an ocean-going vessel designed to transport large quantities of a single product.

Bulk Freight   Freight not in packages or containers.

Bulkhead   A structure inside a cargo tank that prevents cargo from shifting or separating during transit.

Bulkhead Flat   A flat RR car with adjustable bulkheads at each end of car, used for plywood, wallboard, etc.

Bulldozer Company (ICS)   Any bulldozer with a minimum complement of two persons.

Bulletin   Update on the latest development.

Bumping Post (RR)   A braced post placed at end of a stub track to prevent rolling RR cars from going off ends of the rails.

Bunching   The accumulation and tender of RR cars for loading or unloading in excess of orders or contrary to customary schedules.

Bung   A cap or screw-type device that closes the small opening in the top of a metal drum or barrel.

Bunker "B"   A relatively viscous fuel oil (No. 5 fuel) used primarily in marine and industrial boilers.

Bunker "C"   A very viscous fuel oil (No. 6 fuel) used in marine and industrial boilers.

Bureau of Alcohol, Tobacco, and Firearms   Enforces and administers firearms and explosive laws as well as those covering the production, use, and distribution of alcohol and tobacco products.

Bureau of Explosives   An agency of the American Association of Railroads.

Bureau of Solid Waste Management   The branch of the Department of Environmental Resources (DER) responsible for planning, directing, evaluating, coordinating, and organizing a statewide waste management and enforcement program including the Hazardous Waste Management Program.

25

Burning Agent Compound such as gasoline which is used to ignite and sustain combustion of material that would not otherwise burn.

Burning Point Temperature at which a substance sustains combustion; *see* FIRE POINT.

Business Plan A written plan and inventory developed by a business for each facility, site, or branch that provides emergency response guidelines to a release of hazardous materials.

By-Product A material produced in addition to the main product during the manufacturing process. May be a waste or commercial product.

Bypyridyls A group of synthetic organic pesticides that includes the herbicide Paraquat.

C

C Centigrade degrees; also °C.

C4 Four carbons in a chain, e.g., $CH_3CH_2CH_2CH_3$.

C12 Twelve carbons in a chain.

Cab The space in a locomotive "A" unit containing the operating controls and providing shelter and seats for the engine crew.

Cab-Behind Engine Motor truck or truck tractor with driver's compartment and controls located at the rear of a hood-enclosed power plant.

Cab-Over Engine Motor truck or truck tractor with a substantial part of its engine located under the cab.

Cab Signal (RR) A device located in the cab of a locomotive indicating by a display of signals the condition of the track ahead.

Cabin Car Caboose.

Caboose A RR car attached to rear of freight train for office use and living accommodations while in transit.

Caboose Valve (RR) A valve located in the caboose enabling emergency brake applications to be made from rear end of train.

CAER *see* COMMUNITY AWARENESS AND EMERGENCY RESPONSE.

Calcareous Deposit or Growth A growth of calcium carbonate, calcium, or lime on a surface.

Calibrate To measure amount of pesticide that will be applied by the equipment to the target in a given amount of time.

Calibrated Measured.

Calibration (1) A procedure to ensure the accuracy of instrument measurements. (2) Measurement of the delivery rate of application equipment.

California Air Resources Board (CARB) A statewide agency established by the Mulford-Carrell Act of 1967; responsible for coordinating efforts to attain and maintain ambient air quality standards, for conducting research into the causes of and solutions to air pollution, and for regulating motor vehicles to reduce air pollution.

California Department of Forestry and Fire Protection A state agency that protects rural wild lands and other areas not protected by a fire department and/or a fire protection district.

California Environmental Quality Act (CEQA) Enacted in 1970, this state legislation provides a process requiring governmental decision-makers to consider the environmental effects of their decisions and to take feasible measures to prevent significant, avoidable damage to the environment.

California Hazardous Materials Incident Reporting System (CHMIRS) A mandatory post-incident reporting system that collects statistical data on hazardous materials in California.

California Fire Mutual Aid Plan A pre-land agreement made up of all California fire jurisdictions to respond and assist in the event of any incident that has been determined to be outside the local fire jurisdiction's capabilities.

California Law Enforcement Mutual Aid Plan Establishes the state policy for law enforcement mutual aid and outlines the procedures for coordination of alerting, dispatching, and utilizing law enforcement personnel and equipment resources.

California State Emergency Plan The plan established pursuant to Section 8568 of the California Government Code that addresses the state's response to extraordinary emergency situations associated with natural disasters, technological incidents, and war emergency operations.

California Occupational Safety and Health Administration (OSHA) California state agency responsible for enforcement of worker safety laws.

Caller A RR employee who notifies employees to report for duty.

Calorie The amount of heat required to raise the temperature of one gram of water one degree centigrade.

27

Cameo Computer Aided Management of Emergency Operations; a computer storage-retrieval data base of preplanning data for on-scene use at hazardous materials incidents.

Camp (ICS) Site within the general incident area but separate from the base, equipped and staffed to provide food, water, rest, and sanitation services to incident personnel.

Cancelled Pesticide A pesticide use that is no longer registered as a legal use by the EPA; remaining stocks can only be used by order of the Administrator, EPA. (Note: This order is less severe than SUS-PENDED.)

Cancer A group of some 200 diseases grouped together because of their similar growth processes. Each cancer is believed to originate from a single "transformed" cell which does not respond to normal controls over growth, regardless of the part of the body it affects.

Canister A metal or plastic container filled with absorbent materials that filter fumes and vapors from the air before they are breathed in by an applicator.

CANUTEC Canadian Transport Emergency Center; a 24-hour, government-sponsored hot line for chemical emergencies. (Canadian version of CHEMTREC)

Capacity Measure of quantity.

Capacity Indicators Device installed on tank to indicate capacity at specific level; marker.

Capillary Action Absorption.

Capillary Fringe An increasingly moist area that is in continuity with and lies between the saturated zone and the unsaturated zone.

Captive Facilities Facilities located on lands owned by a generator of hazardous waste and operated to provide for the treatment or disposal of that generator's hazardous waste only. *see* ON-SITE FACILITY.

Carboy A large glass or plastic container, usually encased in a protective basket or crate, used to ship hazardous materials, particularly corrosives.

Car Distributor RR employee assigned responsibility of distributing empty freight cars.

Car Dumper (RR) Device for quick unloading of bulk materials such as coal or grain. After being clamped to the rail, the car is tilted or rolled over to discharge the lading.

Car Float Large flat-bottomed boat equipped with tracks on which RR cars are moved in inland waterways.

Car Knocker (slang)   RR car inspector.

Car Lining   Material placed on walls of RR car for protection of cargo.

Car Mile   Movement of a RR car the distance of one mile. Term is used in statistical data.

Car Seals   See SEALS.

Car Stop   A device for stopping motion of a RR car by engaging the wheels, as distinguished from a bumping post which arrests motion upon contact with the coupler of a car; see BUMPING POST.

Carbamate   A synthetic organic pesticide containing carbon, hydrogen, nitrogen, and sulfur; belongs to a group of chemicals that are salts or esters or carbonic acid. Carbamates may be fungicides, herbicides or insecticides. Examples: aldicarb, carbaryl, carbofuran and methomyl.

Carbon Dioxide   A colorless, odorless inert gas that is a byproduct of combustion.

Carbon Monoxide   A highly toxic and flammable gas that is a byproduct of incomplete combustion. Very dangerous even in very low concentrations.

Carbon Sorption   Process whereby activated carbon, known as the sorbent, is used to remove certain wastes from water by preferentially holding them to the carbon surface.

Carbonaceous   Compound having a carbon atom in the chemical structure.

Carcinogen   An agent that produces or is suspected to produce cancer. (FEMA HMCP)

Carcinogenic   The cancer-producing property of a substance or agent; capable of causing cancer.

Card Rack   A small receptacle on the outside of a freight car to receive cards giving shipping directions.

Cardboard   A small board secured to the outside of freight car containing cards giving shipping directions or warning of dangerous lading, etc.

Cardiopulmonary Resuscitation (CPR)   (EMS) Opening and maintaining an airway, providing artificial ventilation, and providing artificial circulation by means of external cardiac compression as defined by the American Heart Association.

Cargo Tank   (Primarily a DOT designation) A container used to transport LP-Gas over the highway as liquid cargo, either mounted

29

on a conventional truck chassis or as an integral part of a transporting vehicle in which the container constitutes in whole or in part the stress member used as a frame. Essentially a permanent part of the transporting vehicle.

Carload   Quantity of freight required for the application of a carload rate for setting transportation fees.

Cartridge   The part of the respirator that absorbs fumes and vapors from the air before the applicator breathes them in.

Cascade System   Several air cylinders attached in series to fill Self–Contained Breathing Apparatus (SCBA) bottles.

C.A.S. Number   Chemical Abstract Service number system, often used in local and state hazardous materials compliance legislation for tracking chemicals in the workplace and in the community. Number assigned to chemicals by the American Chemical Society.

Casing   Steel or plastic tubing that is welded or screwed together to line a borehole.

Casinghead Gas (Oil well gas)   Dissolved gas produced along with crude oil from oil completions.

Casing Pressure   Gas pressure built up between the casing and tubing.

Casualty Collection Point (CCP) (EMS)   Site for the congregation, triage (sorting), preliminary treatment, and evacuation of casualties following a disaster.

Catalyst   A substance which, when present even in small quantities, affects the rate of a chemical reaction but is unchanged itself by the reaction.

Catastrophic Incident   An event that significantly exceeds the resources of a jurisdiction.

Catchment Area (EMS)   Geographic area served by a specified health care facility or EMS agency.

Catenary   A system of wires suspended between poles and bridges supporting overhead contact wires normally energized at 11,000 volts.

Causal Organism   The organism that produces a specific disease.

Caution   A signal word used on pesticide labels to denote slightly toxic pesticides (Toxicity Categories III and IV) as defined by FIFRA (amended).

CBEMA   Computer and Business Equipment Manufacturers Association.

CCMSC   Caribbean Common Market Standards Council.

CDFA   California Department of Food and Agriculture.

Cease and Desist Order   Legal direction to stop any and all activities.

CEC   Commission of the European Community.

CEE   Economic Commission for Europe (UN).

C.E.G.   Certified Engineering Geologist.

Cell   The basic biological unit of plant and animal matter.

Celsius   Temperature scale used by scientists on which water freezes at 0° and boils at 100° at one atmosphere pressure.

CEN   European Committee for Standardization.

CENELEC   European Committee for Electrotechnical Standardization.

Center Dump Car   A RR car that will discharge its entire load between the rails.

Centers for Disease Control (CDC)   A branch of the Public Health Service, Department of Health and Human Services concerned with communicable disease tracking and control.

Center Plate   One of a pair of plates that fit one into the other and support the car body on trucks, allowing them to turn freely under the car. The center pin or king bolt passes through both, but does not really serve as a pivot. The male or body center plate is attached to the underside of the body bolster; the female or truck center plate is attached to the topside of the truck bolster.

Center Sill   The central longitudinal member of the underframe of a RR car which forms the backbone of the underframe and transmits most of the buffing shocks from one end of the car to the other. Freight cars with cushioned underframes use a special type of floating center sill construction.

Centigrade   Celsius.

Centrifugation   A hazardous waste physical treatment process by which heavier particles in the fluid move to the walls of a rotating vessel and are removed.

CEPP   Chemical Emergency Preparedness Program; developed by the EPA to address accidental releases of acutely toxic chemicals.

CEPT   European Conference of Postal and Telecommunications Administrations.

CEQA   California Environmental Quality Act; requires Environmental Impact Reports (EIRs) at site where significant activities occur.

CERCLA   Comprehensive Environmental Response, Compensation and Liability Act; known as CERCLA or the SUPERFUND amendment, addresses hazardous substance releases into the environment and the cleanup of inactive hazardous waste sites. Requires those who release hazardous substances, as defined by the Environmental Protection Agency (EPA), above certain levels (i.e., reportable quantities) to notify the National Response Center.

Cerebrovascular Syndrome   Illness caused by exposure to extremely high doses of ionizing radiation, resulting in severe damage to the central nervous system.

Cerenkov Radiation   An eerie blue glow given off by electrons traveling in a transparent material such as water; visible during operation of some nuclear reactors.

Certification   The recognition by a certifying agency that a person is competent and thus authorized to use or supervise the use of restricted-use pesticides.

Certification Label   A label permanently affixed to forward left side of the trailer stating that the vehicle conforms with all applicable Federal Motor Vehicle Safety Standards in effect on the date of original manufacture.

Certified Applicator   An individual who is certified to use or supervise the use of any restricted-use pesticides covered by his/her certification.

CFR   Code of Federal Regulations.

CGA   Compressed Gas Association.

CGSB   Canadian General Standards Board.

Chain of Custody   A document that tracks samples from the point of collection to delivery at the laboratory. All persons who have physical custody of the samples sign and date acceptance and/or relinquishment.

Chain Reaction   A reaction that stimulates its own repetition.

Charging   see FILLING.

Charles' Law V/T = Constant; the volume of a given mass of gas is directly proportionate to the absolute temperature if the pressure is kept constant.

Chemical Barrier   Chemicals acting as surface tension–modifiers to inhibit the spread of an oil slick on water; short-lived.

Chemical Dispersion   In oil spills, the process of spraying chemical dispersants to remove stranded oil from areas not considered biologically sensitive.

Chemical Manufacturers Association   Parent organization that operates CHEMTREC.

Chemical Name   Scientific name telling the contents or formulae of the active ingredients of the pesticide.

Chemical Oxygen Demand (COD)   A means of measuring the pollution strength of domestic and industrial wastes based upon the fact that all organic compounds, with few exceptions, can be oxidized to carbon dioxide and water by the action of strong oxidizing agents under acid conditions.

Chemical Process   A particular method of manufacturing, usually involving a number of steps or operations.

Chemical Properties   Properties of a material that relate to toxicity, flammability, or chemical reactivity.

Chemical Protective Clothing Material   Any material or combination of materials used in an item of clothing for the purpose of isolating parts of the wearer's body from contact with a hazardous chemical (NFPA 1991, 1–3).

Chemical Protective Suit   Single- or multi-piece garment constructed of chemical protective clothing materials designed and configured to protect the wearer's torso, head, arms, legs, hands, and feet (NFPA 1991, 1–3).

Chemical Resistance   The ability of a material to resist chemical attack. The attack is dependent on the method of test, and its severity is measured by determining the changes in physical properties. Time, temperature, stress, and reagent may all be factors that affect the chemical resistance of a material.

Chemical Resistant Materials   Materials that are specifically designed to inhibit or resist the passage of chemicals into and through the material by the processes of penetration or permeation.

Chemical Sterilants (Chemosterilants)   A chemical that can prevent reproduction in insects (causes sterility).

Chemical Treatment   The processes by which hazardous waste is rendered less hazardous or suitable for transport by changing the chemical composition of such waste. Principal techniques include neutralization, precipitation, ion exchange, chemical dechlorination, and chemical oxidation/reduction.

CHEMNET   A mutual aid network of chemical shippers and contractors; activated when a member shipper cannot respond promptly to

33

an incident involving chemicals. Contact is made through CHEM-
TREC.

CHEMTREC   Chemical Transportation Emergency Center.

**Chemical Hazards Response Information System/Hazards Assess-
ment Computer System (CHRIS/HACS)**   A system developed and
used by the United States Coast Guard. CHRIS manuals contain
chemical-specific data. The HACS is a computerized version of the
CHRIS manual. The system is used by federal on-scene coordinators
during spills and leaks.

Chlorates   Substances that can be herbicides and defoliants; act
as contact poisons, are translocated, and may be absorbed from the
soil to kill both plant roots and tops.

CHLOREP   The Chlorine Emergency Plan; established by the Chlorine Institute to enable the nearest producer of chlorine to respond to
an incident involving chlorine. Contact is made through CHEMTREC.

Chlorinated   A synthetic organic pesticide that contains chlorine,
carbon, and hydrogen; examples: DDT, endrin, lindane.

Chlorinated Hydrocarbon   Chemical in which carbon, chlorine, and
hydrogen are inseparably combined.

Chlorine Kits   Standardized kits commercially manufactured by contract with the Chlorine Institute to provide equipment to control or
stop leaks in chlorine cylinders, tanks, and transportation tank cars.

Chlorosis   The yellowing of a plant's normally green tissue because
of a partial failure of the chlorophyll to develop.

Chlorolysis   Hazardous waste chemical treatment method recycling
chlorinated organic compounds and converting them into useful industrial products through the addition of chlorine.

CHMIRS   The California Hazardous Material Incident Reporting
System; a mandatory post-incident reporting system that collects
statistical data on hazardous material incidents in California. This
data includes a description of the disaster, the location, the time
and date, the state and local agencies responding, the actions taken
by the agencies, and the agency which had primary authority of
responding to the disaster (Chapter 6.95 of the Health and Safety
Code, Title 19 of the California Code of Regulations, and Government Code Section 8574.8(d).k).

Chocolate Mousse   A water-in-oil emulsion containing 50–80%
water.

Cholinesterase   An enzyme (chemical catalyst) found in animals
and humans that helps regulate the activity of nerve impulses and
is necessary for proper nerve function. It is destroyed or damaged

when organic phosphates or carbamates enter the body in any path or in any fashion.

Cholinesterase Inhibitor   Any carbamate, organophosphate, or other pesticide that can interrupt the action of enzymes.

CHRIS/HACS   see CHEMICAL HAZARDS RESPONSE INFORMATION SYSTEM/HAZARD ASSESSMENT COMPUTER SYSTEM.

Chronic Health Effects   Long-term effects from a one-time or a repeated exposure to a substance.

Chronic Poisoning   Poisoning that occurs as a result of repeated exposures over a period of time.

Chronic Toxicity   Small, repeated doses of a poisonous substance absorbed or ingested by animal or person over a period of time.

Circuitous Route   An extremely indirect route.

Circus Loading (RR)   Method of loading highway trailers by moving them over the ends of the cars.

CIS   A term used in cases of Cis-Trans Stereoisomerism to indicate that certain elements or groups are on the same side of the molecule, such as Cis-Form or Trans-Form.

CISD   Critical Incident Stress Debriefing; stress reduction processes designed to address the special needs of emergency response personnel in dealing with situations that cause strong emotional reactions or interfere with the ability to function.

Clandestine   Something that is secret, hidden, or concealed.

Clandestine Laboratory   An operation consisting of a sufficient combination of apparatus and chemicals that either could have been or was used in the illegal manufacture/synthesis of controlled substance.

Clarifier   A gravity apparatus for removing settled solids from a fluid.

Class 1 Disposal Sites   Areas where complete protection is provided for all time for the quality of ground and surface waters from all wastes deposited therein, and where protection against hazard to public health and wildlife resources is provided.

Class II Disposal Sites   Class II disposal sites are those at which protection is provided to water quality from Group 2 and Group 3 wastes. Class II-1 sites are those overlying usable ground water where geologic conditions either are naturally capable of preventing lateral and vertical hydraulic continuity between liquids and gases emanating from the waste in the site and usable surface or ground waters, or have been modified to achieve such capability.

The following criteria must be met to qualify a disposal site as Class II:

(1) Disposal areas shall be protected by natural or artificial features so as to assure protection from any washout and from inundation which could occur as a result of tides or floods having a predicted frequency of once in 100 years.

(2) Surface drainage from tributary areas shall not contact Group 2 wastes in the site during disposal operations and for the active life of the site.

(3) Gases and leachate emanating from waste in the site shall not unreasonably affect ground water during the active life of the site.

(4) Subsurface flow into the site and the depth at which water-soluble materials are placed shall be controlled during construction and operation of the site to minimize leachate production and assure that the Group 2 waste materials will be above the highest anticipated elevation of the capillary fringe of the ground water. Discharge from the site shall be subject to waste discharge requirements.

Classification (1) To arrange or sort into uniform categories or classes, usually by size, weight, color, organic/inorganic, etc. (2) (RR freight cars) A destination and routing code used on switch lists for ease in switching cars.

Classification of Explosives

CLASS A EXPLOSIVE—A material or device presenting a maximum hazard through detonation.

CLASS B EXPLOSIVE—A material or device presenting a flammable hazard and functions by deflagration.

CLASS C EXPLOSIVE—A material or device containing a restricted quantity of either Class A or Class B explosives or both, but presents a minimum hazard.

Classification of Fires

CLASS A FIRE—Fire in combustibles and flammable gases or liquids.

CLASS B FIRE—Fire in combustible solids, such as wood, paper, or cardboard.

CLASS C FIRE—Fire involving energized electrical equipment.

CLASS D FIRE—Fire involving combustible metals such as magnesium, zirconium, etc.

36

Classification—Pesticides   The process of assigning pesticides into groups according to common characteristics; may be by target species, chemical nature, manner of formulation, mode of action or toxicity.

Classification—Poisons

CLASS A POISON—A poisonous gas or liquid of such nature that a very small amount of the product is dangerous to life.

CLASS B POISON—U.S. Dept. of Transportation category for highly toxic material, a substance known to be so toxic to human life that it affords a severe health hazard during transportation.

Classification Yard   A yard where RR cars are grouped according to their destinations and made ready for proper train movement.

Class Rate   A rate based on an assigned class rating (a percentage of first class) published in the Uniform Freight Classification.

Clean Air Act   A set of national standards for ambient air quality; define the principal types of pollution and the levels of each that should not be exceeded. Requires states to draw up "state implementation plans" for achieving the ambient air standards in each air quality control region in the state.

Clean Room   A room designed to maintain a defined level of air cleanness under operating conditions.

Cleanout Fitting   A fitting installed in the top of a tank to facilitate washing of the tank interior.

Cleanup   Incident scene activities directed at the removal of hazardous materials, contamination, debris, damaged containers, tools, dirt, water, and road surfaces in accordance with proper and legal standards, and returning the site to as near as normal as it existed prior to the incident (Sacramento Hazardous Materials Team).

Cleanup Company (Hazardous Waste)   A commercial hazardous waste hauler, which must meet specific state requirements, that is available for hire to specifically remove, transport, and dispose of hazardous wastes.

Cleanup Operation   An operation where hazardous substances are removed, contained, incinerated, neutralized, stabilized, cleared up, or in any other manner processed or handled with the ultimate goal of making the site safer for people or the environment (CFR 29 Section 1910.120(a)(3) [SARA]).

Clean Water Act (CWA)   Federal legislation aimed at protecting the nation's water and setting state water quality standards for interstate navigable waters as the sole basis for pollution control and enforcement. The main objective is to restore and maintain the chemical, physical, and biological integrity of the nation's waters.

Clear Board (RR)   A signal indication displayed to advise that no train orders are being held.

Clearance Lamps   Lamps mounted on the permanent structure of a vehicle to indicate overall width of the vehicle, one on each side of the vertical centerline, at the same height, and as near the top as practicable; may be mounted on the front or rear of the vehicle.

Clearance or Clearance Limit (RR)   The limiting dimensions of height and width for cars in order that they may safely clear all bridges, tunnels, station platforms, and other structures as well as equipment on adjacent tracks.

Cleat   A strip of wood or metal used to afford additional strength, to prevent warping, or to hold in position.

Clinical   Pertaining to the symptoms and cause of disease.

Closed Portion   Any portion of a facility where hazardous waste treatment, storage, or disposal operations have been closed.

Closure   Actions taken by the owner or operator of a hazardous waste facility to prepare the site for long-term care and to make it suitable for other uses after wastes are no longer accepted.

CMA   Chemical Manufacturers Association.

CNS   Central Nervous System.

Coal Car (RR)   A car for carrying coal, usually a hopper car.

Cobalt Bomb   Theoretically, a nuclear weapon encased in cobalt, which would produce large amounts of the highly penetrating and long-lasting gamma radiation emitted by cobalt 60.

Code of Federal Regulations (CFR)   The formal name given to those books or documents containing specific regulations provided for by laws adopted by the U.S. Government.

COFC (RR)   Container-on-flatcar.

Coffin   A thick-walled container (usually lead) used for transporting radioactive materials.

Cohesion   The attraction of like molecules.

Coke Rack (RR)   A slatted frame or box applied above the sides and ends of gondola or hopper cars to increase their cubic capacity for the purpose of carrying coke or other freight in which the bulk is large relative to the weight.

Cold Zone   *see* SUPPORT ZONE.

Coliwassa   A tool designed to provide stratified sampling of a liquid container.

Collection   The act of picking up waste materials at homes, businesses, or industrial sites, and hauling it to a facility for further processing, transfer to large vehicles, or disposal.

Collection Center   A facility designed for accepting materials from individuals, usually for recycling.

Colorimetric Tubes   Glass tubes containing a chemically treated substrate that reacts with specific airborne chemicals to produce a distinctive color. The tubes are calibrated to indicate approximate concentrations, as air.

Combination Stop and Tail Lamps   An electrical device serving the dual purpose of a stop lamp and tail lamp.

Combustibility or Combustible   A material that can, under most conditions, enter into the combustion process; burnable.

Combustible Liquids   Liquids with a flash point above 140°F according to NFPA definition.

Combustible Metals   Metals that will burn.

Combustible Solids   Those materials that ignite relatively easily and are subject to rapid flame propagation at moderate temperature.

Combustion   Fire; an exothermic chemical reaction due to rapid oxidation of a fuel, involving light and heat.

Combustion Explosion   Sudden fracture of a container or structure accompanied by a shock wave (sound) caused by the overpressure created by the expansion of a gas (often mainly air) within the container or structure in response to absorption of heat produced by combustion of a flammable mixture within the structure.

Combustion Product   Material generated during the burning or oxidation of matter.

Command (ICS)   The act of directing, ordering, and/or controlling resources (personnel and/or material) by virtue of explicitly legal or delegated authority.

Command Post   Central control point for an incident, located at a safe distance from an accident site where the coordinator, responders, and technical representatives can make response decisions, deploy manpower and equipment, maintain liaisons with media, and handle communications.

Command Staff   (ICS) Personnel who report directly to incident commander, usually include information officer, safety officer, and liaison officer.

Commercial Waste   Waste material originating in wholesale, retail, or service establishments such as office buildings, stores, hotels, universities, and warehouses.

Commodity Rate   A rate applicable to a specific commodity between certain specified points.

Common Exposure Route   A likely way (oral, dermal, respiratory) that a pesticide may reach and/or enter an organism.

Common Name   A well-known, simple name of a pesticide accepted by the Pesticide Regulation Division of the EPA.

Common Name of Pesticide   Well-known made-up name accepted by the EPA to identify the active ingredients in a pesticide.

Communicable Disease   A disease that can be transmitted from one person to another; also known as contagious disease.

Communications Center   The location within a general incident area equipped and staffed to provide and control communications for an emergency situation.

Communications Officer   Individual on command post staff responsible for flow of information and maintenance of the communication system at the incident.

Communications System   A collection of individual communication networks, transmissions systems, and relay, control, and base stations capable of interconnection and interoperations designed to form an integral whole. The individual components must be technically compatible, employ common procedures, and operate in unison.

Community Awareness and Emergency Response (CAER)   A program developed by the Chemical Manufacturers Association (CMA) to provide guidance for chemical plant managers to assist them in taking the initiative in cooperating with local communities to develop integrated (community/industry) hazardous materials response plans.

Community Right-to-Know   Legislation requiring business establishments to provide chemical inventory information to local agencies or the public.

Compactor   Any power-driven mechanical equipment designed to compress and reduce the volume of waste materials.

Compactor Truck   A large truck with enclosed body having special power-driven equipment for loading and compressing waste materials.

Company (Fire Usage)   Any piece of fire response equipment having

40

a full complement of personnel [National Incident Management System (NIIMS)].

Compartment Tank Cars (RR)   A tank car with the body divided into several sections for the purpose of carrying smaller shipments or different commodities in each compartment.

Compartmentizer Car (RR)   A box car equipped with movable bulkheads which can be used to divide the car into separate compartments.

Compatibility   The mixture of two or more pesticides without their effectiveness being reduced or altered.

Compatibility Charts   Permeation and penetration data supplied by manufacturers of chemical protective clothing to indicate chemical resistance and breakthrough time of various garment materials as tested against a battery of chemicals. This test data should be in accordance with ASTM test practices and NFPA standards.

Compatible   Capable of mixing two or more compounds without affecting properties of the other.

Compensation   Payments awarded either through the courts or through a government-administered fund to cover injury or damage caused by exposure to hazardous substances. Awards usually cover lost income, out-of-pocket medical expenses, and pain and suffering.

Competency-Based Curriculum (EMS)   A program wherein specific objectives are defined for each of the separate skills and successful completion of an examination demonstrating mastery of such skills in required.

Complexing Agent   A material that forms a chemical complex with a second material (very tightly bound at the molecular level) when the two come in contact.

Compost   Decomposed organic matter, often used for soil enrichment.

Composting   Controlled method of decomposing organic matter by the natural activity of microorganisms.

Compound   A substance consisting of two or more elements that have been united chemically.

Compressed Gas   Any material or mixture having in the container an absolute pressure exceeding 40 psia at 70°F, or regardless of the pressure at 70°F, having an absolute pressure exceeding 104 psia at 130°F.

Compressed Gas Association (CGA)   Association of firms producing and distributing compressed, liquefied and cryogenic gases; also manufacturers of related equipment. Submits recommendations to appropriate government agencies to improve safety standards and methods of handling, transporting, and storing gases; acts as advisor to regulatory authorities and other agencies concerned with safe handling of compressed gases; collaborates with national organizations to develop specifications and standards of safety.

Compressed Gas in Solution A non-liquefied gas dissolved in a solvent, at high pressures.

Compton Effect The glancing collision of a gamma ray with an electron. The gamma ray gives up part of its energy to the electron.

Concealed Damage Damage to the contents of a package which appears externally to be in good order.

Concentrate A pesticide formulation (liquid or dry) as it is sold before diluting; usually contains a high percentage of active ingredient to save shipping and storage charges and yet be of convenient strength and composition for dilution.

Concentration Amount of active ingredient contained in a given volume or weight.

Concentration Limits The level for each hazardous waste constituent which triggers initiation of a corrective action program.

Condensate Hydrocarbons that are in the gaseous state under reservoir conditions but which become liquid either in passage up the hole or at the surface.

Condition A particular state of being of a person, thing, or set of circumstances.

Condition Use Permit (CUP) A discretionary permit issued by cities and counties; required for certain projects; imposes conditions designed to assure that the project is compatible with the local General Plan and zoning ordinances and that impacts to neighboring land uses are minimized.

Conduction A method of heat transfer whereby molecules of the substance transfer heat from one to another by direct contact.

Conductor Train service employee in charge of train or yard crew.

Cone of Depression A cone-shaped depression formed in a water table when ground water is removed.

Confidence Level or Confidence Limit (95%) A level of data reliability achieved by setting a percent confidence limit. A ninety-five percent confidence limit is the limit of the range of analytical

42

values within which a single analysis will be included ninety-five percent of the time.

Confined Aquifer An aquifer whose upper and/or lower boundaries are confined by an impermeable geologic formation, e.g., a clay layer; an aquifer in which ground water is under pressure, e.g., artesian conditions.

Confinement Those procedures taken to keep a material in a defined or localized area.

Confining Layer An aquitard or impermeable layer that confines the limits of an aquifer.

Conflicting Routes Two or more routes over which simultaneous movements of cars cannot be made without possibility of collision.

Connate Water Water inherent to the oil-producing formulation, or fossil sea water trapped in the pore spaces of sediments during their deposition.

Connecting Carrier A railroad that has a direct physical connection with another or forming a connecting link between two or more railroads.

Connection Box Contains fittings for trailer emergency and service brake connections and an electrical connector to which the lines from the towing vehicle may be connected. Also called junction box, light box, bird box.

Consignee The person who is to receive a shipment; the addressee to whom the item is being shipped.

Consignee Marks A symbol placed on packages for export, generally consisting of a square, triangle, diamond, circle, cross, etc. with designated letters and/or numbers for the purpose of identification.

Consignor Person or firm from whom shipment originates; also called shipper.

Consist A rail shipping paper similar to a cargo manifest. May contain a list of cars in the train in order or a list of those cars carrying hazardous materials and their location on the train.

Constructive Placement When, due to some disability on the part of the consignor or consignee, a RR car cannot be placed for loading or unloading, it is considered to be under constructive placement and subject to demurrage rules and charges, as if it were actually placed.

Consultant Any licensed engineering firm involved in the assessment of cleanup of an underground tank leak. The consultant is hired by the responsible party.

43

Contact   Being exposed to an undesirable or unknown substance that may pose a threat to health safety.

Contact Herbicide   An herbicide that kills primarily by contact with plant tissue rather than as a result of translocation. Only the portion of a plant contacted is directly affected.

Contact Poison   A pesticide that kills when it touches or is touched by a pest.

Container   Any portable device in which material can be stored, handled, transported, treated, or disposed of.

Container (Explosives)   The type of article/material/substance in which explosive/incendiary/chemical elements are placed for the purpose of constituting a device.

Container (Freight)   An article of transport equipment that is (1) of a permanent character and strong enough for repeated use; (2) specifically designed to facilitate the carriage of goods by one or more modes of transport without intermediate reloading; (3) fitted with devices permitting its ready handling, particularly its transfer from one mode to another. This definition does not include vehicles. Also called freight container, cargo container, intermodal container.

Container Appurtenances   Items connected to container openings needed to make a container a gastight entity. These include safety relief devices, shutoff, backflow check, excess flow check, internal valves, liquid level gauges, pressure gauges, and plugs.

Container Assembly   An assembly consisting essentially of the container and fittings for all container openings. These include shutoff valves, excess flow valves, liquid level gauging devices, safety relief devices, and protective housing.

Container Car   A flat or open top RR car such as a gondola, on which containers of freight are loaded.

Container Chassis   A trailer chassis having simply a frame with locking devices for securing and transporting a container as a wheeled vehicle.

Container Specification Number   Number found on a shipping container preceded by the initials DOT, which indicates the container has been constructed to federal specifications.

Containership   A ship specially equipped to transport large freight containers in horizontal or, more commonly, in vertical container cells. The containers are loaded and unloaded by special cranes.

Containment   All activities necessary to bring the incident to a point of stabilization and to establish a degree of safety for emergency personnel greater than that which existed upon arrival.

Containment Shell   A gastight enclosure around a nuclear reactor or other nuclear facility designed to prevent fission products from escaping to the atmosphere.

Contaminant   A substance or process that poses a threat to life, health, or the environment; a toxic material found as a residue in or on a substance where it is not wanted.

Contaminate   To add an unwanted material, such as a pesticide, that may do harm or damage; to pollute, make impure, or make unfit for use.

Contaminated-Exhaust System   An air-cleaning system designed to remove harmful or potentially harmful particulates, mists, or gases from the air exhausted from an operating area.

Contamination   The introduction of an unwanted substance into any material or the air. Radioactive contamination is harmful to people and the environment; contaminants affect fire behavior when introduced to normally stable fuels.

Contamination Control Line   The established line around the contamination reduction zone that separates it from the support zone.

Contamination Reduction Zone   Term used by United States Coast Guard to identify an area of moderate hazard, where contamination threat or spread to the immediate surrounding area is low: the area immediately outside of the inner hot zone; see also WARM ZONE.

Contingency Plan   A document presenting an organized and coordinated plan of action to limit potential pollution in the event of fire, explosion, or discharge of hazardous materials; defines specific responsibilities and tasks.

Continuous Air Monitor   An instrument that detects airborne particulate and gaseous radioactivity and sounds an alarm at specified concentrations.

Continuous Seals (RR)   A RR term denoting that the seals on a car have remained intact during the movement of the car from point of origin to destination; or, if broken in transit, that seals were broken by proper authority and without opportunity for loss to occur before new seals were applied. see SEALS.

Control   The procedures, techniques, and methods used in the mitigation of a hazardous materials incident, including containment, extinguishment, and confinement.

Control Agents   Any material used to contain or extinguish a hazardous material or its vapors.

45

Control Zones   The designation of areas at a hazardous materials incident based upon safety and the degree of hazard (NFPA 472, section 1–3). *see also* SUPPORT ZONE, WARM ZONE, HOT ZONE and DECONTAMINATION CORRIDOR.

Controlled Area   Any area controlled by the user for purposes of radiation safety.

Controlled Point   A location where the signals and/or switches of a RR traffic control system are operated and/or controlled from a distant location by a train dispatcher.

Controlled Siding   A siding governed by signals under the control of a train dispatcher or operator.

Convection   Method of heat transfer whereby the heated molecules circulate through the medium (gas or liquid).

Converter Dolly   An auxiliary undercarriage assembly with a fifth wheel and towbar used to convert a semitrailer to a full trailer; also called a dolly or converter gear.

Coordination   Systematic analysis of a situation and available resources, developing relevant information to inform appropriate command authority of viable alternatives for the most effective actions to meet specific objectives.

Cooperating Agency (ICS)   An agency supplying assistance other than direct suppression, rescue, support, or service functions to the incident control effort.

COPANT   Pan American Standards Commission.

Corner Fittings   Strong metal devices located at the corners of a container, having several apertures which normally provide the means for handling, stacking, and securing the freight container.

Corner Structures   Vertical frame components located at the corners of a container, integral with the corner fittings.

Cornered   RR term to denote that a car has been struck by another RR car because it was not in the clear.

Corrective Actions   Actions taken by the incident commander to correct the problem at hand in a hazardous materials emergency.

Corrective Action Measures   The removal or treatment in place of any hazardous constituents that exceed concentration limits in the ground water below a land disposal facility.

Corrosive   The quality of a substance that causes the gradual deterioration of another material by chemical processes, such as oxidation or attack by acids.

46

Corrosive Material   Any liquid or solid that can destroy human skin tissue; a liquid that has severe corrosion rate on steel.

Corrosive Poison   A type of poison containing a strong acid or base which will severely burn the skin, mouth, stomach, etc.

Corrosives   Substances that attack and wear away another substance by strong chemical action.

Corrugated Paper   A rigid structural paper shaped in parallel furrows and ridges.

COS   Corporation for Open Systems.

Cosmic Radiation   Radiation originating in outer space.

Cost Recovery   A procedure that allows for the agency having jurisdiction to pursue reimbursement for all costs associated with a hazardous materials incident.

Coulomb   In physics, a unit of electrical charge equivalent to one ampere second.

Council on Environmental Alternatives (CEA)   Encourages people to conserve, rather than consume, their environment. Concentrates on the areas of energy and provides specific recommendations to encourage individuals to recognize and assume responsibility for environmentally sound choices available to them.

Counter   A device measuring ionizations (counts) of particles per unit of time.

Counts Per Minute (CPM)   Number of ionizations of particles measured per minute. *see* ACTIVITY, DECAY.

County Solid Waste Management Plan (CoSWMP)   A plan that sets forth a comprehensive program for solid waste management pursuant to Government Code Section 66780.

Coupler (RR)   An appliance for connecting RR cars or locomotives. Government regulations require that these must couple in an automatic manner on impact and must be uncoupled without a person having to move between cars.

Coupler Centering Device (RR)   A device for maintaining the coupler normally in the center line of draft but allowing it to move to either side when a car is rounding a curve while coupled to another car.

Coupler Knuckle Lock (RR)   The block that drops into position when the knuckle closes and holds it in place, preventing uncoupling.

47

Coupler Lock Lifter (RR) The mechanism inside the coupler head that, when moved by the uncoupling rod, lifts the knuckle lock so that the knuckle may open.

Coupon Small metal strip exposed to corrosive systems in the oil industry for the purpose of determining nature and severity of corrosion.

Covered Gondolas (RR) Gondola equipped with removable cover to protect lading from weather exposure in transit.

Covered Hopper Car (RR) A hopper car with a permanent roof, roof hatches, and bottom openings for unloading.

CP (RR slang) see CONSTRUCTIVE PLACEMENT or CONTROLLED POINT.

CPCFA California Pollution Control Financing Authority; state agency that helps businesses finance pollution control devices.

CPG 1–3 Federal Assistance Handbook: Emergency Management, Direction and Control Programs, prepared by FEMA. Provides states with guidance on administrative and programmatic requirements associated with FEMA funds.

CPG 1–5 Objectives for Local Emergency Management, prepared by FEMA. Describes and explains functional objectives that represent a comprehensive and integrated emergency management program. Includes recommended activities for each objective.

CPG 1–8 Guide for Development of State and Local Emergency Operations Plans, prepared by FEMA see EOP.

CPG 1–8A Guide for the Review of State and Local Emergency Operations Plans, prepared by FEMA. Provides FEMA staff with a standard instrument for assessing EOPs that are developed to satisfy the eligibility requirement for receiving Emergency Management Assistance funding.

CPG 1–35 Hazard Identification, Capability Assessment, and Multi-Year Development Plan for Local Governments, prepared by FEMA. As a planning tool, it can guide local jurisdictions through a logical sequence for identifying hazards, assessing capabilities, setting priorities and scheduling activities to improve capability over time.

Crack a Valve Open a valve slightly so that it leaks just a little to relieve pressure on a line.

Cradle-to-Grave The tracking of the source, quantity, concentration, and type of hazardous waste from generation through final disposal.

Crew A group of individuals working together as a unit.

Crib Portion of ballast between two adjacent RR ties.

Cripple (RR slang)   *see* BAD ORDER.

Critical   The point at which some quality, property, or phenomenon undergoes change.

Critical Assembly   The assembly of sufficient fissionable material and a moderator to sustain a fission chain reaction at a very low power level, permitting study of the behavior of the components for various fissionable materials in different geometrical arrangements.

Critical Mass   Minimum amount of fissionable material capable of supporting a nuclear chain reaction.

Critical Pressure   Pressure at which a gas becomes a liquid.

Critical Temperature   Temperature at which a gas becomes a liquid.

Critical Velocity   Lowest water current velocity that will cause loss of oil under the skirt of a containment boom.

Criticality   The state of sustaining a chain reaction, as in a nuclear reactor.

Cross Tie (RR)   Transverse member of RR track structure to which rails are spiked to provide proper gauge and to cushion, distribute, and transmit the stresses of traffic through the ballast to the road-bed.

Crossing   A RR structure used where one track crosses another at grade, consisting of four connected frogs.

Crossmember   Transverse members in an underframe.

Crossmember Spacing   Distance between crossmember centers.

Crossover   Two turnouts with RR track between, connecting two nearby and usually parallel tracks.

Crossover Line   A tank piping system that allows unloading from either side of tank.

Crude Oil   Petroleum in its natural form before being subjected to any refining process.

Crummy   Caboose.

Cryogenic   Pertaining to liquified gases stored at temperatures approaching absolute zero. Normally, they have a boiling point of about −100°C.

Cryogens   Gases that must be cooled to a very low temperature in order to change from a gas to a liquid.

CSA   Canadian Standards Association.

CSMA   Chemical Specialties Manufacturers Association.

Cu. Yds.   Cubic yards.

Cubic Capacity   Useful internal load-carrying space, usually expressed in cubic feet, yards, or gallons.

Cubical Capacity   Carrying capacity of a RR car according to measurement in cubic feet.

Cullet   Crushed glass used in glass-making. Use of cullet speeds melting of silica sand, reducing the energy needed in the manufacturing process.

Cumulative Effect   The result of some poisons that build up or are stored in the body so that small amounts contacted over a period of time can sicken or kill a person or animal.

Cupola   Small cabin built on roof of a caboose to afford a means of lookout for the train crew.

Curbside   Side of a trailer nearest the curb when trailer is traveling in a normal forward direction (right-hand side); opposite of roadside.

Curie   The basic unit used to describe the intensity of radioactivity in a sample of material; equal to thirty-seven billion disintegrations per second, approximately the rate of decay of one gram of radium.

Current of Traffic   Movement of trains on a track in a designated direction specified in the timetable.

Cushion Underframe   A term designating the framework of a railway car designed to prevent shocks and impact stresses from damaging the car structure or its lading.

Customary   A system of measurement used in the United States. Units of the customary system are feet, yards, ounces, pounds, and various units of volume (cups, pints, quarts, gallons).

Cut   (1) (Verb). To uncouple a RR car. (2) (Noun). A group of RR cars coupled together. (3) (Noun). That part of the right-of-way excavated out of a hill instead of running over it or tunneling through it.

Cut Oil   Oil containing water; also called wet oil.

Cutout Cock (Air brake)   A valve that will bypass or cut out the brake system for a car when closed. Closing this valve does not interfere with braking operation of other cars in the train.

Cyclotron   A particle accelerator in which charged particles receive synchronized accelerations by electrical fields as the particles spiral outward from their source. The particles are kept in the spiral by a powerful magnetic field.

Cylinder   A portable container constructed to DOT cylinder specifications or, in some cases, constructed in accordance with the ASME Code of a similar size and for similar service. Maximum size permitted under DOT specs is 1,000 pounds water capacity.

D

Damage   Harm to an inanimate object that disrupts its intended functional continuity.

Damage Assessment   A gathering of information on the type, extent, and costs of damage after an incident.

Damage Free Car   RR car equipped with special bracing devices to decrease the possibility of damage to lading; DF Car.

Damming   A procedure of constructing a dike or embankment to totally immobilize a flowing waterway contaminated with a liquid or solid hazardous substance (EPA, 600/2-77-277).

Danger   Signal word used on pesticide labels to denote high toxicity (Toxicity Category I) as defined by FIFRA (amended); always accompanied by skull and crossbones symbol and the word *Poison*.

Dangerous Cargo Manifest   Manifest used on ships containing a list of all hazardous materials on board and their location on the ship.

Dangerous When Wet   A label required for water-reactive materials (solid) being shipped under US DOT, ICAO, and IMO regulations. Any of this labeled material that is in contact with water or moisture may produce flammable gases. In some cases, these gases may spontaneously combust (49 CFR 171.8).

Daughter   Nuclide formed by the radioactive decay of another nuclide, which in this context is called the parent.

Deadhead   RR Train and/or engine crew moved without performing service from one terminal to another at railroad convenience and for which they are paid.

Deadman   A buried timber, log, or beam designed to serve as an anchorage for a guy wire or cable that supports a structure such as a wood or steel column, derrick, or mast.

Deadman Control (RR)   A foot pedal or brake valve that must be kept in a depressed position while the locomotive is operating. A release from this position initiates an air brake application after a short delay.

Dead Rail   A second set of RR tracks over a scale used when cars are not being weighed.

Dead Well   An oil well that will not flow.

Debilitating Illness or Injury   An illness or injury that temporarily or permanently prevents an employee of an organization from engaging in normal duties and activities.

Debris   Any material resulting from the demolition or destruction of any structure, including stones, bricks, rocks, concrete, gravel, or earth.

Debug   To detect, locate, and remove mistakes from a routine or malfunctions from a computer.

Decay (Radioactive)   The spontaneous transformation of one nuclide into a different nuclide or into a different energy state of the same nuclide. The process results in a decrease, over time, in the number of the original radioactive atoms in a sample. It involves the emission from the nucleus of alpha particles, beta particles (or electrons), or gamma rays; the nuclear capture or ejection of orbital electrons; or fission. Radioactive disintegration.

Decay Product   A nuclide, either radioactive or stable, resulting from the disintegration of a radioactive material.

Decay Rate   The decrease in activity of a radioactive material within a given time; usually expressed in terms of the period during which half of the atoms will disintegrate (i.e., half-life).

Dechlorination   An experimental hazardous waste chemical treatment process that produces a change in the carbon-chlorine bond in organic compounds high in chlorine (e.g., PCB, kepone) through the use of reducing agents.

Declared Emergency   An action taken by a jurisdiction according to the California Emergency Services Act and local ordinances in response to the impact of a real or threatened hazard that exceeds local resources.

Decomposition   A change in the composition of organic matter to a less complex form; may be accomplished by introduction of heat, through addition of neutralized chemicals, or through biodegradation.

Decomposition Gases   Vapors produced by the decomposition of garbage or organic matter; some are combustible, such as methane.

Decomposition Product   Material produced by the physical or chemical degradation of a parent material.

Decon Popular abbreviation referring to the process of decontamination.

Decontaminate To make safe, purify, or make fit for use again by removal of any unsafe substances, such as pesticides or radioactivity.

Decontamination The physical and/or chemical process of absorbing, destroying, neutralizing, or making harmless biological, chemical, or radioactive substances to make a person, object, or area safe.

Decontamination Corridor A corridor that acts as a protective buffer and bridges between the hot zone and the cold zone; located within the warm zone, where decontamination stations and personnel are located to apply decontamination procedures.

Decontamination Officer (ICS) An officer appointed by the incident commander to identify location of the decontamination corridor, assign stations, manage all decontamination procedures, and identify types of decontamination necessary.

Decontamination Team (Decon Team) Group of personnel and resources operating within a decontamination corridor.

Deep-Well Injection The disposal of hazardous wastes by pumping them into deep wells so that they can percolate through porous or permeable subsurface rock and be contained within surrounding layers of impermeable rock or clay. Used extensively for disposal of oil field wastes.

Defect Card (RR) A record-keeping device employed by train crews to report problems with the cars in a train. The document is forwarded to the closest repair track for resolution of the problem.

Defect Card Receptacle Small metal container in which defect cards are placed; located underneath the RR car for protection from the weather.

Definitive Care (EMS) Medical treatment that should remedy an illness or injury.

Deflagration An exothermic reaction in a material which propagates from the burning gases to the unreacted material by conduction, convection, and radiation. In this process the reaction zone progresses at a rate less than the velocity of sound.

Deflocculating Agent An adjuvant that prevents precipitation or settling of solids in the suspension fluid.

Defoliant A pesticide causing the leaves of a plant to drop off.

Defoliate To strip of leaves; to apply a defoliant.

Degradability   Ability of a chemical to break down into less complex compounds or elements.

Degradation   The breakdown of a more complex chemical into a less complex form; can be a result of the action of microbes, water, air, sunlight, or other agents. (2) When referring to protective clothing, the loss of physical properties of the item due to exposure to chemicals, actual wear and tear, or the ambient conditions in the environment.

Degrade   To decompose; break down.

De-inking   A process in which most of the ink, filler, and other extraneous material is removed from printed waste paper, producing pulp which can be used with varying percentages of virgin material in the manufacture of new paper.

Delayed Toxic Exposure Effect   The conditions in which symptoms of an exposure are not present immediately after the exposure, but instead are delayed for a relatively short period of time; *see* CHRONIC EFFECT.

Deleterious Substances   Substances not normally harmful to humans that may be harmful to the environment.

Delisting a Waste   The process by which a particular facility proves to DER or EPA that waste from that facility is fundamentally different from waste listed as hazardous and does not meet any of the criteria under which the hazardous substance was listed. The waste product may be excluded from hazardous waste regulation if it does not meet any of the criteria of a hazardous waste.

Delivering Carrier   The railroad that delivers a shipment to the consignee.

Delta   Designates a difference in something, as in ΔP for difference in pressure.

Deluge Set   A device used in fire suppression to apply large volumes of water (master streams) to a fire.

Demurrage   A penalty charge assessed by railroads for the detention of cars by shippers or receivers of freight beyond a specified free time.

Density   (1)   The ratio of the weight of an object or mass to its volume. (2) Number of resident and non-resident persons per square mile.

Department of Commerce (DOC)   A federal agency with a primary mission to encourage, serve, and promote economic development and technological advancement.

Department of Defense (DOD)   Provides the military forces needed to deter war and protect the security of our country.

Department of Energy (DOE)   Provides the framework for a comprehensive and balanced national energy plan through the coordination and administration of the energy functions of the federal government; is responsible for long-term, high-risk research, development, and demonstration of energy technology, the marketing of federal power, energy conservation, the nuclear weapons program, regulation of energy production and use, and a central energy data collection and analysis program.

Department of Fish and Game   Enforces provisions of the state fish and game code, which prohibits pollution of waters, including ocean waters.

Department of Health Services (DHS)   A California agency created in 1978 to administer health care delivery programs, health protection programs, and publicly financed health programs. In the area of toxics this department carries out its health protection function through the following program elements:

- environmental health hazard assessment
- toxic substances control
- sanitary engineering
- radiologic health
- food and drug
- local environmental health
- laboratory services

Department of Justice (DOJ)   Serves as counsel for the citizens of the nation; represents them in enforcing the law in the public interest; through its thousands of lawyers, investigators, and agents, it plays a key role in ensuring healthy competition of business in our free enterprise system, in safeguarding the consumer, and in enforcing drug, immigration and naturalization laws; plays a significant role in protecting citizens through its efforts for effective law enforcement, crime prevention, crime detection, and prosecution and rehabilitation of offenders; conducts all suits in the Supreme Court in which the United States is concerned; represents the government in legal matters.

Department of Labor (DOL)   Purpose is to foster and develop the welfare of the wage earners of the United States, improve their working conditions, and advance their opportunities for profitable employment.

Department of State (DOS)   Advises the President in formulation and execution of foreign policy; promotes long-range security and well-being of the United States; determines and analyzes the facts relating to American overseas interest; makes recommendations on

55

policy and future action; takes necessary steps to carry out established policy; engages in continuous consultation with the American public, the Congress, other U.S. departments and agencies, and foreign governments.

Department of Transportation (DOT) Assures the coordinated, effective administration of the transportation programs of the federal government and develops national transportation policies and programs conducive to the provision of fast, safe, efficient, and convenient transportation at the lowest cost consistent therewith.

Depleted Uranium Uranium having a smaller percentage of uranium 235 than the 0.7% found in natural uranium; obtained from the spent or used fuel elements as by-product tails or residues of uranium isotope separation.

Depressed Center Flat Car A RR flat car with the section of the floor between the trucks depressed to permit loading of high shipments within overhead clearance limits; a well flat car.

Derail A RR track safety device designed to guide a car off the rails at a selected spot as a means of protection against collisions or other accidents; commonly used on spurs or sidings to prevent cars from fouling the main line.

Dermal Through or by the skin; of or pertaining to the skin.

Dermal Toxicity The degree to which a poison is absorbed through the skin.

Designated Facility (EMS) A hospital established by local authority to perform specified emergency medical services systems functions.

Dessicant A pesticide used to draw moisture from a plant, plant part, or insect. Used primarily for pre-harvest drying of actively growing plant tissues when seed or other plant parts are developed but only partially mature, or for drying plants that normally do not shed their leaves, such as rice, corn, small grains, and cereals.

Dessication Dehydration (removal of tissue moisture) by chemical or physical action.

Destination The place to which a RR shipment is consigned.

Detailed Survey The measurement and evaluation of radiation hazards in a given area.

Detection Monitoring Program Procedure utilized to ensure discovery of any leakage from a land treatment facility.

Detectors The following serve in detection monitoring programs:

COMBUSTIBLE GAS INDICATOR (CGI) DETECTOR — Measures the concentration of a combustible gas or vapor in air.

CORROSIVITY (pH) DETECTOR—A meter or paper that indicates relative acidity or alkalinity; generally uses an international scale of 0 (acid) through 14 (alkali-caustic); see pH.

FLAME IONIZATION (FDI) DETECTOR—A device used to determine the number of carbon atoms in a molecule.

GAS CHROMATOGRAPH/MASS SPECTROMETER DETECTOR—An instrument used for analyzing organics.

HEAT DETECTOR—An instrument used to detect heat by sensing infrared waves.

RADIATION BETA SURVEY DETECTOR—An instrument used to detect beta radiation.

RADIATION DOSEMETER DETECTOR—An instrument that measures the amount of radiation to which a person has been exposed.

RADIATION GAMMA SURVEY DETECTOR—An instrument used for the detection of ionizing radiation, principally gamma radiation, by means of a gas-filled tube.

TEMPERATURE DETECTOR—An instrument, either mechanical or electronic, used to determine temperature of ambient air, liquids, or surfaces.

Detonation (1) An exothermic reaction characterized by the presence of a shock wave in the material that establishes and maintains the reaction, propagating at a rate equal to or exceeding the speed of sound. (2) A hazardous waste chemical treatment method that treats explosive waste by rapid combustion. (3) Explosive destruction.

Detoxify To make harmless; to take out, take away, or neutralize a poison; to remove a poisonous effect.

Detritus Loose material resulting from rock disintegration or abrasion; suspended material in the water column including fragments of decomposing flora and fauna, and fecal pellets produced by zooplankton and associated bacterial communities.

Deuterium An isotope of hydrogen whose nucleus contains one neutron and one proton, making it about twice as heavy as the nucleus of normal hydrogen, which is only a single proton. Often called heavy hydrogen, it occurs in nature as one atom to 6,500 atoms of normal hydrogen and is not radioactive.

DHS Department of Health Services (State).

di- A prefix indicating the number two; also bi.

Dialysis Process of separating a mixture of substances in solution by using a membrane as a filtering agent. Substances move through the membrane at varying rates and separate according to their relative molecular weights.

Diamond   RR crossing.

-diene   A suffix indicating the unsaturation of two ethylenic (two double bond) linkage, as in a diolefin such as pentadiene.

Diesel Electric Locomotive   A train engine in which one or more diesel engines drive electric generators, which in turn supply electric motors (usually series D.C.), which are geared to the driving axles.

Diffusion   (1) Mixing of substances, usually gases and liquids, due to molecular motion. (2) The spreading out of a substance to fill a space.

Diffusion Plant (Gaseous)   A method of isotopic separation based on the fact that gas atoms or molecules with different masses will diffuse through a porous barrier (or membrane) at different rates. Method is used by the AEC to separate uranium 235 from uranium 238; requires large gaseous-diffusion plants and enormous amounts of electric power.

Dike   An embankment of natural or man-made materials constructed to contain or obstruct the movement of liquids, sludges, or other substances.

Dike Overflow   A dike constructed in such a fashion that allows uncontaminated water to flow unobstructed over the dike while keeping the contaminant behind the dike.

Dike Underflow   A dike constructed in such a fashion that allows for the uncontaminated water to flow unobstructed under the dike while keeping the contaminant behind the dike.

Diluent   Any liquid or solid material used to weaken or carry an active ingredient such as a concentrated pesticide.

Dilute   To weaken a solution by adding water, oil, or other liquid or solid.

DIN   Deutsches Institut für Normung (Germany).

Dinky   A small RR engine used around roundhouses or backshops for switching.

-dioc   A suffic indicating a dicarboxylic acid, such as pentandioc acid.

Dip   The complete or partial immersion of a plant, animal, or object in a pesticide.

Dip Tube   Device for pressure unloading of a product through the top of a tank.

Direct-Cycle Reactor System A nuclear power plant system in which the coolant or heat transfer fluid circulates first through the reactor and then directly to a turbine.

Direct Disease Transmission Transmission of a communicable disease from one person to another due to direct contact with infected blood, body fluids, or other infectious materials.

Direct Gas-Fired Tank Heater A device that applies heat from a gas burner flame directly to a portion of a container surface in contact with LP-Gas liquid in order to vaporize the product at the rate needed to supply the connected gas-consuming devices.

Directed Application Aiming a pesticide at a specific area such as at a portion of a plant, animal, or structure or at a row or bed.

Direction and Control (Emergency Management) The provision of overall operational control and/or coordination of emergency operations at each level of the Statewide Emergency Organization, whether it be the actual direction of field forces or the coordination of joint efforts of governmental and private agencies in supporting such operations.

Disaster A perilous condition that exceeds a jurisdiction's capability to control with immediately available resources.

Disaster Assistance Center (DAC) A facility jointly established by the federal and state coordinating officers within or adjacent to a disaster–impacted area to provide disaster victims "one-stop" service in meeting their emergency and/or rehabilitation needs. It will usually be staffed by representatives of local, state, and federal governmental agencies, private service organizations, and certain representatives of the private sector.

Disaster Field Office (DFO) A central facility established by the federal coordinating officer within or immediately adjacent to disaster-impacted areas to be utilized as a point of coordination and control for state and federal governmental efforts to support disaster relief and recovery operations.

Disaster Medical Services (EMS) Medical services provided to disaster victims that minimize morbidity and mortality.

Disaster Operations Center (DOC) Location for the coordination of disaster response activities by a state department of health services.

Disaster Response Component (EMS) Resources and arrangements necessary to adequately respond to mass casualty situations on a local, regional, or statewide basis.

Disaster Service Worker Includes public employees and any unregistered person impressed into service during a state of emergency or local emergency by a person having authority to command the aid of citizens. Does not include any member registered as an active firefighting member of any regularly organized volunteer fire department, having official recognition, and full or partial support of the county, city, town, or district in which such a fire department is located.

Disaster Support Area (DSA) A designated area on the periphery of an emergency incident from which relief resources can be received, stored, allocated, and dispatched.

Disaster Welfare Inquiry (DWI) A service that provides health and welfare reports about relatives and certain other individuals believed to be in a disaster area and when disaster-caused dislocation or disruption of normal communications facilities precludes normal communications.

Discharge An intentional or accidental spilling, leaking, pumping, pouring, dumping, emitting, or any other release of hazardous waste, hazardous waste constituents, or hazardous materials which, when released into land or water, become hazardous waste.

Discretionary Project An undertaking requiring the exercise of judgment when the public agency decided to approve or disapprove a particular activity; as distinguished from situations where that agency merely has to determine if there has been conformity with applicable statutes, ordinances, or regulations.

Discriminator An electronic circuit that selects signal pulses according to their pulse height or voltage. Used to delete extraneous radiation counts or background radiation or as the basis for energy spectrum analysis.

Disease An alteration of a person's health, with a characteristic set of symptoms, that may affect the entire body or specific organs. Diseases have a variety of causes and are known as infectious diseases when caused by a pathogenic microorganism such as a bacteria, virus, or fungus.

Disinfectant Pesticide that controls germs.

Disinfection Effective killing by chemical or physical processes or procedures of virtually all recognized pathogenic microorganisms on inanimate objects capable of causing infectious disease. Some microbial forms such as bacterial endospores can survive disinfection; see STERILIZATION.

Disintegration A nuclear transformation; to break or decompose into constituent elements.

Dispatch (ICS) Implementation of a decision to move resources from one place to another.

Dispensing Device (Dispenser) A device normally used to transfer and measure LP-Gas for engine fuel into a fuel container, serving the same purpose for an LP-Gas service station as that served by a gasoline dispenser in a gasoline service station.

Dispersants Chemicals that reduce the surface tension between oil and water, thereby facilitating the breakup and dispersal of an oil slick throughout the water column in the form of an oil-in-water emulsion. Chemical dispersants can only be used in areas where adverse biological damage will not occur and then only when approved for use by government regulatory agencies.

Dispersing Agent An adjuvant that reduces the attraction between particles.

Dispersion The distribution of spilled oil into the upper layers of the water column by natural wave action or application of chemical dispersants. In shoreline cleanup and restoration, the removal of stranded oil through wave action (natural dispersion), application of chemical dispersants, or use of one of various hydraulic dispersion techniques.

Dispersion Mechanism The process by which a material scatters in one or more directions.

Disposable Designed to be used for a brief period of time, then discarded; neither durable nor repairable.

Disposal Drum A specially constructed drum used to overpack damaged or leaking containers of hazardous materials for shipment.

Disposal Site The location where any final deposition of hazardous waste occurs. (Section 66052, Title 22, California Administrative Code; the Department of Health Services is proposing to revise this definition.)

Disposal Well An oil well through which water (usually salt water) is returned to subsurface formations.

Dissolution The process of dissolving one substance in another. In oil spills, a process contributing to the weathering of spilled oil whereby certain slightly soluble hydrocarbons and various mineral salts present in oil are dissolved in the surrounding water.

Dissolved Product The water-soluble fuel components, namely, benzene, toluene, and xylene.

Distributing Plant (LP-Gas) A facility that receives LP-Gas in tank car, truck transport, or truck lots, distributing this gas to the end user by portable container delivery, tank truck, or through gas piping. Such plants have bulk storage (2,000 gallons water capacity or more) and usually have container filling and truck loading facilities on the premises; bulk plant. Normally no persons other than the plant management or plant employees have access to these facilities.

Distributing Point (LP-Gas) A facility other than a distributing plant normally receiving LP-gas by tank truck, which fills small containers or the engine fuel tanks of motor vehicles on the premises. Any such facility having LP-Gas storage of 100 gallons or more water capacity and to which persons other than the owner of the facility or his employees have access, is considered a distributing point; an LP-Gas service station.

Diversion The intentional movement of a hazardous material in a controlled manner so as to relocate it in an area where it will pose less harm to the community and the environment.

Division (1) (RR) Portion of the railroad assigned to the supervision of a superintendent, usually consisting of yards, stations, and sidings. (2) (ICS) The organizational level within the incident command system having a responsibility for operations within a specific geographical area. A division officer usually commands a series of fire companies and reports directly to the operations chief.

DOE United States Department of Energy.

DOHS State Department of Health Services; state agency whose responsibilities include toxic waste management and regulation in California.

DOLLY see CONVERTER DOLLY.

Dome The circular fixture on top of a tank car containing valves and relief devices.

Dormant Spray A pesticide application made before trees and other plant life begin to leaf out in the spring.

Dose (1) The amount of ionizing radiation energy absorbed per unit mass of irradiated material at a specific location, such as a part of the human body, measured in rems or, in an inanimate body, measured in rads. (2) Amount, quantity, or portion of a pesticide that is applied to a target. (3) A consistent measure used in testing to determine acute and chronic toxicities.

Dose Equivalent Amount of effective radiation when modifying factors have been considered. The product of absorbed dose multiplied by a quality factor multiplied by a distribution factor; expressed numerically in rems.

Dose Projections A calculated estimate of the potential dose to individuals at a given location; determined by the quantity of radioactive material released from the source and the appropriate meteorological transport and dispersion parameters.

Dose Rate The radiation dose delivered per unit time; measured in rems per hour.

Dosimeter   Radiation-measuring device, such as a lapel badge or an ionization chamber.

Dosimeter Charger   Device utilized to induce a static electric charge on a dosimeter.

DOT   United States Department of Transportation.

Double   (1) Two consecutive tours of duty. (2) Putting train together when part of train is on one track and balance on another.

Doubles   Trailer combination consisting of a truck tractor, semi-trailer, and a full trailer coupled together.

Double Containment   Arrangement of double barriers whereby the second barrier provides backup protection against leakage through, or failure of, the first.

Double Deck (Stock car)   A RR car with a second floor (often removable) halfway between the ordinary floor and roof to increase carrying capacity of car for small livestock.

Double Filtration   (1) Arrangement of double barriers in which the second provides protection against leakage of the first. (2) A series arrangement intended to increase total filtration efficiency.

Double Gloving   The practice of wearing an additional set of gloves over those already in place.

Double Walled Tank   A container with two complete shells which provide both primary and secondary containment. The outer shell must provide structural support and must be constructed primarily of nonearthen materials.

Doughnut   A ring of wedges supporting a string of pipe, or a threaded, tapered ring used for same purpose.

Downwind   Direction toward which the prevailing wind is blowing.

Draft Gear   Unit on a RR car that forms connection between the coupler rigging and the center sill and cushions force of impact of shocks incidental to train movements.

Drag   (1) A train of empty RR cars. (2) A heavily laden train. (3) Group of cars for movement from one point to another within a terminal.

Drainage Well   A well installed to drain water at or near ground surface.

Drawbar   Coupler.

Drawhead   Head of an automatic coupler, exclusive of the knuckle, knuckle pin, and lock.

Dresser Sleeve   A slip-type collar used to join plain-end pipe.

Drift   Movement of particles by wind and air currents from the target area to an area not intended to be treated.

Drill Track (RR)   Track connecting with the ladder track, over which engines and cars move in switching.

Drilling (Car)   Handling or switching of RR cars in freight yards.

Drinking Water Supply   Any raw or finished water source that is or may be used as a public water system or as drinking water by one or more individuals.

Drome Box   On tractor-trailer vehicles, a large boxlike structure between the cab and the trailer, often containing hazardous materials.

Drop   RR switching movement when cars are cut off from an engine and allowed to roll free into a track.

Drop-Axle   An axle in which the centerline of the beam portion is offset from the centerline of the spindles.

Drop Bottom Car (RR)   A gondola car with a level floor, equipped with drop doors for discharging the load.

Drop Brake Shaft (RR)   A brake shaft for flat cars that normally extends above the floor, but can be dropped down should conditions of the lading require.

Drop-End Gondola Car (RR)   Gondola car with end doors that can be dropped for shipping long material extending over more than one car.

Drop Frame   A two-level section of trailer providing proper coupler height on the forward end and a lower floor height for the remainder of the trailer length.

Droplet Breakaway   A type of boom failure resulting from excessive current velocity; the head wave formed upstream of the mass contained within a boom becomes unstable.

Dry Gas   (1) Natural gas produced without liquids. (2) A gas that has been treated to remove all liquids.

Dry Well (Dry Hole)   A well that does not extend into the water table or saturated zone.

DS   Dansk Standardiseringsraad (Denmark).

Dummy Coupler   A dust cap fitting used to seal the opening in an air brake hose connection when not in use.

Dump   An open land site where waste materials are disposed of in a manner that does not protect the environment, is susceptible to open burning, or is exposed to the elements, vermin, and/or scavengers.

Dump Car   A RR car capable of discharging load either through doors or by tipping the car body.

Dunnage   The material used to protect or support freight in or on cars, such as bracings, false floors, meat racks, etc.

Durable   Long-lasting in spite of hard wear or frequent use.

Dust   A finely ground dry mixture combining a small amount of pesticide with an inert carrier such as clay, talc, or volcanic ash.

Dutchman   (1) A piece of pipe that has been twisted off inside a female connection. (2) A short section of material, such as belting or pipe, used to lengthen existing equipment.

Dynamic Braking   Using the motors of a locomotive as generators and dissipating power through resistors to control train speed, after which air brakes are utilized to bring the train to a full stop.

Dzus Fastener   Trade name for a screw-like fastener with a slotted head used on aircraft. One-half turn either way fastens or releases the fastener.

E

Easy Sign   Hand signal indicating the train or other vehicle is to move slowly.

EBS   see EMERGENCY BROADCASTING SYSTEM.

Ecology   Study of the relationship between a plant or animal and its surroundings.

Economic Poison   Term formerly used for pesticide.

Ecosystem   Interacting system of a biological community and its non-living environment.

EDB   Ethylene dibromide.

EDC   Ethylene dichloride.

Edema   A swelling of body tissues as a result of fluid retention.

EEC   European Economic Community.

Effluent   (1) Discharge or outflow of water from ground or subsurface storage. (2) Solid, liquid, or gas wastes that enter the environment as a by-product of man-oriented processes.

EFTA   European Free Trade Association.

EIA   Electronic Industries Association.

Electric Locomotive   A locomotive that receives power from an overhead contact wire or third rail and uses that power to drive electric motors connected by gears to the driving axles.

Electrically Locked Switch   Hand-operated switch equipped with an electrically controlled device that restricts the movement of the switch.

Electrically Neutral   Term given to an object with a net electrical charge of zero; example: an atom of hydrogen has one proton $(+1)$ and one electron (-1); the two opposite charges cancel, and the atom is neutral.

Electrodialysis   Process of separating substances in a solution by dialysis using an electric field as the driving force.

Electrolysis   A hazardous waste chemical treatment method by which chemical changes are accomplished at the surface of electrodes immersed in a chemical solution and carrying an electric current.

Electrolyte   (1) A group of chemicals essential to biological functions. In solution, the molecules dissociate into ions and aid in transferring electrical impulses in the nervous system. (2) A substance that, when dissolved in a suitable solvent or when fused, becomes an ionic conductor.

Electromagnetic Radiation   Radiation consisting of associated and interacting electric and magnetic waves that travel at the speed of light.

Electron   A component of an atom; travels in a distant orbit around a nucleus.

Electrostatic Precipitator   Device on smokestacks collecting small particles of dust by giving them an electrical charge and attracting them to a collecting electrode.

Element   (1) The most simple substance that cannot be separated into more simple parts by ordinary means. There are approximately 100 elements. (2) A group of atoms which display the same physical and chemical properties and have the same number of protons in their nuclei. (3) (ICS) Any identified part of the Incident Command System organizational structure.

Elementary Particles   The simplest particles of matter and radiation.

66

Elevated Release   A release of radioactive effluents via the ventilation release point on a reactor containment building, including effluents from containment building and a supplementary leak collection and release system.

ELISA   Enzyme-Linked Immunosorbent Assay; a test used to detect antibodies to the AIDS virus indicating infection.

ELOT   Hellenic Organization for Standardization (Greece).

Embargo   Order prohibiting acceptance and/or handling of freight at certain points or via certain routes due to emergencies, congestion, strikes, etc.

Emergency   (1) Common use—A situation of concern that has developed suddenly and unexpectedly, and that demands prompt action. (2) Generic definition—A disaster situation or condition of extreme peril to life and/or property, resulting from other than war or labor controversy, which is or is likely to be beyond local capability to control without assistance from other political entities. (3) Federal definition—Any hurricane, tornado, storm, flood, high-water, wind-driven water, tidal wave, tsunami, earthquake, volcanic eruption, landslide, mudslide, snowstorm, drought, fire, explosion, or other catastrophe in any part of the United States that requires federal emergency assistance to supplement state and local efforts to save lives and protect public health and safety or to avert or lessen the threat of a major disaster.

Emergency Actions   The assessment, corrective, and protective actions taken during the course of an emergency.

Emergency Application   A quick, heavy reduction of brake pipe pressure made when a train must be stopped in the minimum distance possible.

Emergency Brake Valve   A valve for applying the train brakes in an emergency, connected to the brake pipe by a branch pipe and operated by releasing brake pipe air to the atmosphere.

Emergency Broadcast System (EBS)   A system that enables authorities to communicate with the general public through commercial broadcast stations during time of disaster. EBS uses facilities and personnel of the broadcast industry on a voluntary organized basis. It is operated by the industry under rules and regulations of the Federal Communications Commission.

Emergency Communications Center   A facility designated by a political jurisdiction as responsible for receiving and transmitting emergency communications.

Emergency Coordinator   The individual in an organization who has been appointed in a preplan to be responsible for direction of on-site actions during an emergency.

Emergency Dose Limit   A level of projected absorbed dose following a nuclear incident above which the total risk to health of an individual is considered excessive.

Emergency Implementing Procedures   Preplanned detailed actions to be taken in case of emergency incident.

Emergency Medical Care   The provision of treatment to patients, including both basic (BLS) and advanced life support (ALS) and other medical procedures prior to the arrival at a hospital or other medical facility.

Emergency Medical Services (EMS)   A group, department, or agency that is trained and equipped to respond in an organized manner to emergency situations where there is a potential need for pre-hospital emergency medical care.

Emergency Operating Center (EOC)   A facility used for the centralized direction and/or coordination of emergency operations; the actual protected site where government officials exercise centralized direction and control in an emergency. An effective EOC must provide adequate working space and be properly equipped to accommodate its staff, have a capability to communicate with field units and other EOCs, and provide protection commensurate with the projected risk at its location. Additionally, the EOC normally serves as a resource center and coordination point for additional field assistance. It also provides executive directives and liaison to state and federal government, and considers and mandates protective actions.

Emergency Operations Plan   A document that identifies the available personnel, equipment, facilities, supplies, and other resources in the jurisdiction, and states the method or scheme for coordinated actions to be taken by individuals and government services in the event of natural, man–made, and attack-related disasters.

Emergency Period   A period that begins with the recognition of an existing, developing, or impending situation that poses a potential threat to a community. It includes the warning period and impact and recovery phases. An emergency period is not considered over until the immediate and ensuing effects of the threat no longer constitute a hazard to life and property.

Emergency Plans   Official and approved documents describing the authority, principles, policies, concepts of operations, methods, and procedures to be applied to carry out emergency operations or render mutual aid during emergencies.

Emergency Planning Levels (EPL)   A determination of specific levels of risk to acute chemical exposure from airborne hazardous materials discharges. An EPL is usually established for a minimum

68

of a one-hour exposure. Different levels can be established for low or high risk individuals. EPLs are chemical-specific. There are four EPLs, as follows:

LEVEL I—The safe exposure level. At this level or below adverse health effects are not expected for about 99 percent of the population.

TRANSITION RANGE—A range where there are adverse effects for most of the population, but more importantly, for a smaller number of sensitive individuals, this range can cause serious effects. For these individuals exposure at this level can create Level II injuries.

LEVEL II—The serious injury level. Exposure at this level for more than one hour may not be reversible without medical intervention. Effects can include impairment of one's judgement and reduced ability to take appropriate health protection measures.

LEVEL III—The fatality level. Severe irreversible injury can occur. Complete incapacitation and the inability to escape or take precautionary measures may occur.

Emergency Public Information (EPI) Information disseminated to the public at large by official sources during an emergency period, using both electronic and printed media outlets. May include information regarding evacuation, survival, health preservation, what not to do, and status information on the emergency.

Emergency Public Information System The network of public information officers (PIOs) and staff personnel who operate emergency public information centers.

Emergency Reserve Account for Hazardous Material Incidents A fund administered by the California Department of Health Services to finance actions only for the purpose of remediation or prevention of threats of fire, explosion, or human health hazards resulting from a release or potential release of a hazardous substance (California Health and Safety Code 25354).

Emergency Response Response to any occurrence that has or could result in a release of a hazardous substance (29 CFR 1910).

Emergency Response Organization An organization that employs emergency response personnel.

Emergency Response Personnel Personnel assigned to organizations that have the responsibility for responding to different types of emergency situations (NFPA 1991, 1–3).

EMI The Emergency Management Institute; a component of FEMA's National Emergency Training Center located in Emmitsburg, Maryland. It conducts resident and nonresident training activities for federal, state, and local government officials, managers in the private economic sector, and members of professional and volunteer organizations on subjects that range from civil nuclear preparedness systems to domestic emergencies caused by natural and technological hazards. Nonresident training activities are also conducted by state emergency management training offices under cooperative agreements that offer financial and technical assistance to establish annual training programs that fulfill emergency management training requirements in communities throughout the nation.

Eminent Domain   The right of a government to appropriate private property for necessary public use, with compensation paid to the landowner.

Empty Packaging   Any packaging having a capacity of 110 gallons or less that contains only the residue of a hazardous material in Table 2 of 49 CFR, section 172.504.

EMS Agency   A local or regional emergency medical services provider.

EMS   Emergency Medical Services; functions as required to provide urgent medical care to ill or injured persons by trained EMS providers.

Endothermic   A process or chemical reaction accompanied by absorption of heat.

-ene   A suffix indicating the unsaturation of one ethylenic, as in pentene. It is often used as a suffix in naming aromatic hydrocarbons, such as benzene, toluene, and napthalene.

Enforcement Agency   The agency designated by the authority having jurisdiction to enforce provisions of local, state, or federal statutes.

Engine   Any emergency response vehicle providing specified levels of pumping, water, hose capacity, and personnel.

Enriched Material   Material in which the percentage of a given isotope present in a material has been artificially increased so that it is higher than the percentage of that isotope naturally found in the material.

Enteric Precautions   A system of precautions to prevent the transmission of disease by the oral/fecal route.

Entomologist   Scientist specializing in the study of insects.

Entry Point   A specified and controlled access into the hot zone at a hazardous materials incident.

Entry Team Leader   The person responsible for the overall entry operations of assigned personnel within the hot zone. (FIRESCOPE ICS-HM)

Environment   (1) The air, water, and the earth, sometimes called the biosphere. (2) The sum total of all external conditions that may act upon an organism or community to influence its development or existence.

Environmental Impact Report (EIR)   A detailed statement describing and analyzing the significant environmental effects of a project and discussing ways to mitigate those effects.

Environmental Fate   What happens to a chemical once it is released or escapes into the environment.

Environmental Protection Agency (EPA)   A federal agency charged with implementing the Resource Conservation and Recovery Act and having general responsibility for administering programs that address the environmental problems of water and air pollution, toxic substances, pesticides, radiation, noise, and solid waste management.

Environmental Sensitivity   The susceptibility of a local environment or area to any disturbance that might decrease its stability or result in either short- or long-term adverse impacts. Environmental sensitivity generally includes physical, biological, and socioeconomic parameters.

Environmentally Sensitive Area   Areas exceptionally responsive to environmental change and especially prone to irreversible ecological upset. These can include wetlands, floodplains, permafrost areas, critical habitats of endangered species, and recharge areas of aquifers.

EOC Liaison   Person designated to establish communications between the incident scene and the EOC.

EOP   Explosive Ordinance Disposal; military or civil bomb squads.

EOQC   European Conference of Postal and Telecommunications Administrations.

EP Toxicity   A characteristic indicating the likelihood that certain constituents could be leached by a medium in significant concentrations, as determined by a specific extraction procedure.

EPA (or U.S. EPA)   United States Environmental Protection Agency.

EPA Registration Number   Number appearing on a pesticide label to identify the individual pesticide product; may appear as "EPA REG NO."

Equalizer Beam   Suspension device used to transfer and maintain equal load distribution between two or more axles of suspension; rocker beam.

Equalizer Hanger Bracket for mounting equalizer beam of multiple-axle suspension to trailer frame, which allows for beam's pivotal movement; center hanger.

Eradicant A pesticide that kills the pest after it appears.

Eradicant Fungicide Pesticide that kills a fungus after it appears on or in a plant.

ERT Environmental Response Team; a group of highly specialized experts available through EPA twenty-four hours a day.

Erythema Abnormal redness of the skin, as in inflammation.

Erythrocytes Red blood cells.

ESA European Space Agency.

Estuary Widened channel of the mouth of a river in which influences of the tides are felt; delicate ecosystems serving as nurseries, spawning and feeding grounds for large groups of marine life, and provide shelter and food for birds and wildlife.

Etiologic Agent A living microorganism that may cause human disease; germ.

ETSI European Telecommunications Standards Institute.

EURATUM European Atomic Energy Community.

Evacuation Removal of residents from an area of danger.

Evacuation Area Area surrounding an emergency incident in which personnel and/or residents may be subject to ready removal.

Evacuee An individual who is moved from or leaves a hazard area to a lesser hazard area with the anticipation of return when the hazard has abated.

Evaporate To become a gas and disappear into the air; to vaporize.

Evaporation (1) The process of a solid or liquid turning into a gas. (2) A hazardous waste physical treatment process by which suspended and dissolved solids are separated from liquid waste.

EWOS European Workshop on Open Systems.

Exclusion Zone see HOT ZONE.

Exothermic A process or chemical reaction accompanied by heat.

Exploratory Boreholes Any temporary excavation for determining

72

subsurface conditions; includes vapor probes, hydropunch holes, geotechnical borings, etc.

Explosion   Sudden release of large amount of energy in a destructive manner; a result of powders, mists, or gases undergoing instantaneous ignition, liquids or solids undergoing sudden decomposition, or a pressurized vessel undergoing overpressure rupture; accompanied by such force as to generate tremendous heat, cause severe structural damage, and occasionally generate a shock wave and/or propel shrapnel.

Explosion-Proof Equipment   An instrument whose enclosure is designed and constructed to prevent the ignition of an explosive atmosphere. Certification for explosion-proof performance is subject to compliance to ASTM standards.

Explosive   A term applied to solid or liquid substances possessing the faculty under certain circumstances of undergoing instantaneous decomposition, extending throughout their entire mass. The process is accompanied by a considerable disengagement of heat, the substance being wholly or partially converted into gaseous products.

EXPLOSIVE CLASS A—Any of nine types of explosives as defined in Title 49 CFR 173.53. A material that when detonated creates a shock wave which travels faster than the speed of sound.

EXPLOSIVE CLASS B—Those explosives which in general function by rapid combustion rather than by detonation; include some explosive devices such as special fireworks, flash powders, some pyrotechnic signal devices, and liquid or solid propellant explosives which include some smokeless powders (49 CFR 173.88).

EXPLOSIVE CLASS C—Certain types of manufactured articles which contain Class A and/or Class B explosives, as components but in restricted quantities; certain types of fireworks; small arms ammunition (49 CFR 173.100).

Explosive Range   The boundary lines of mixture or concentration of flammable vapor and air, which, if ignited by an outside heat source, will propagate flame; see FLAMMABLE RANGE.

Export   To send goods to a foreign country.

Exposure   Contact of living tissue with ionizing radiation. see CONTAMINATION.

Exposure Determination   The review of job classifications within a work setting that results in employees being divided into two groups. The first group includes job classifications in which all employees have occupational exposure. The second group includes job classifications in which some employees have occupational exposure.

73

Exposure Incident   A specific eye, mouth, other mucous membrane, broken skin or parenteral contact with blood or other potentially infectious materials that results from the performance of an employee's duties; an example: puncture from a contaminated sharp.

Exposure Rate   The rate of exposure to ionizing radiation, measured in roentgens/hour.

Extended Planning Zone   An area beyond the emergency zone and out to approximately twenty miles in which extensive planning (but not evacuation) for public protection is accomplished.

Extra Train   Train not included in a timetable schedule.

Extremely Hazardous Materials   Uses this term for chemicals that must be reported to appropriate authorities if released above the threshold reporting quantity. The list of these substances and their threshold reporting quantities is identified in Title 40 CFR, part 355. Releases of extremely hazardous substances as defined by EPA must be reported to the National Response Center.

Extremely Flammable   A liquid that has a flash point of 20°F or lower, determined by closed cup of Seta flash test.

F

F   Fahrenheit degrees.

Fabrication in Transit   The stopping of steel products at a point located between the points of origin and destination for further process or manufacture; for example: steel beams to be fabricated as bridge girders.

Facility   The on-site structures and all adjoining land and rights of way that are used for testing, storing, or disposing of hazardous wastes.

Facing Movement   The movement of a train over the points of a switch that face in a direction opposite to that in which the train is moving.

Faecal (Fecal) Pellets   Solid or semi-solid excretion products (faeces or feces) that are enclosed within a thin membrane, such as produced by zooplankton and some other invertebrates.

Failure Mode   The way in which a failure occurred.

Fallout   Airborne particles containing radioactive material that fall to the ground following a nuclear explosion.

Fatigue   Failure of a metal under repeated loading.

Fault-Tree Analysis   A means of analyzing hazards. Hazardous events are first identified by other techniques such as HAZOP. Then all combinations of individual failures that can lead to that hazardous event are shown in the logical format of the fault tree. By estimating the individual failure probabilities, and then using the appropriate arithmetical expressions, the top event frequency can be calculated.

Fauna   Animals in general or animal life as distinguished from plant life (flora); usually used in reference to all the animal life characteristic of or inhabiting a particular region or locality.

FDA   United States Food and Drug Administration

Feasibility Study   Detailed examination of the technical, environmental, engineering, economic, legal, and practical suitability of a proposed facility or technology for use at a specific location.

Federal Agency   (Federal definition) Any department, independent establishment, government corporation, or other agency of the executive branch of the federal government, including the United States Postal Service but excluding the American Red Cross.

Federal Coordinating Officer (FCO)   (Federal definition) The person appointed by the president of the United States to coordinate federal assistance following an emergency or major disaster declaration.

Federal Disaster Assistance   Assistance to disaster victims or state or local government provided by the federal government in the form of monetary or in-kind assistance in accordance with federal statutes.

Federal Insecticide, Fungicide, and Rodenticide Act (FIFRA) (amended)   An act that required pesticides to be registered and labelled, made it illegal to detach or destroy pesticide labels, and provided for pesticide inspections. FIFRA now requires EPA to determine whether a pesticide "will perform its intended function without causing unreasonable adverse effects on the environment" or human health.

Federal Water Pollution Control Act (1972) (FWPCA)   *see* CLEAN WATER ACT.

Feedback   An element of a system that returns a portion of the output to the input, thus allowing the system to evaluate itself.

Feeding in Transit   The stopping of shipments of livestock at a point located between the points of origin and destination.

FEMA   United States Federal Emergency Management Agency.

FEMA-REP-5   Guidance for Developing State and Local Radio-

75

logical Emergency Response Plans and Preparedness for Transportation Accidents, prepared by FEMA. Provides a basis for state and local governments to develop emergency plans and improve emergency preparedness for transportation accidents involving radioactive materials.

Ferrous Metals   Metals that are predominantly composed of iron; most are magnetic.

FHWA   Federal Highway Administration; DOT division concerned with highway construction and usage. Other divisions of DOT relate to air, rail, and water transportation.

Fibrosis   A condition marked by an increase of interstitial fibrous tissue.

FID   Flame Ionization Detector.

FIEI   Farm and Industrial Equipment Institute.

Field   An area consisting of a single oil reservoir or multiple reservoirs all grouped on, or related to, the same individual geological structural feature and/or stratigraphic condition.

Field Administrative Limit   A pre-established limit on radiation absorbed dose for emergency personnel. The limit is set by local jurisdictions for use in managing field actions in response to radiological incidents.

Field Blank   A sample container filled with organic free water that is taken on the field trip. It is opened and exposed at the sampling site to detect contamination from air exposure. The water sample may be poured into appropriate containers to simulate actual sampling conditions. Contamination from air exposure can vary considerably from site to site, therefore the need for this sample should be evaluated relative to the sample situation. Reference material (i.e., chemically defined soil) can be used in lieu of organic free water as dictated by the sampling needs.

Field Duplicate   A second field sample collected identically to and immediately after the first sample; provides a measure of analytical precision and second sample confirmation; provides a means of determining random error when adequate numbers of duplicates are collected. Field duplicates may also be collected as splits and can also serve as blind field samples.

FIFRA (amended)   see FEDERAL INSECTICIDE, FUNGICIDE, and RODENTICIDE ACT (amended).

Fifth Wheel   A device used to connect a truck tractor or converter dolly to a semitrailer in order to permit articulation between the units. It generally is composed of a lower part consisting of a trunnion, plate, and latching mechanism mounted on the truck tractor (or dolly), and a kingpin assembly mounted on the semitrailer.

Fifth Wheel Pickup Ramp   A steel plate designed to lift the front end of a semitrailer to facilitate engagement of kingpin into fifth wheel.

Fill Opening   Opening in top of tank used to load container, usually incorporated in manhole cover.

Fill, Filling   Transferring liquid LP-Gas into a container.

Filling by Weight   see WEIGHT FILLING.

Film Badge   A light-tight package of photographic film worn like a badge by workers in nuclear industry or research and used to measure possible exposure to ionizing radiation. The absorbed dose can be calculated by the degree of film darkening caused by the irradiation.

Filter   A device or substance for straining out solid particles or impurities from a liquid or gas.

Filter Bank   A parallel arrangement of filters on a common mounting frame enclosed within a single housing.

Filter Canister   A container filled with sorbents and catalysts that remove gases and vapors from air drawn through the unit. The canister may also contain an aerosol (particulate) filter to remove solid or liquid particles. Air-purifying canister-type breathing apparatuses are not approved for use during emergencies by the fire service in California.

Filter Pack   Sand or gravel placed in the annular space of the well between the borehole and the screened or perforated interval of the well casing; usually washed and graded material; see GRAVEL PACK ENVELOPE.

Filtration   A hazardous waste physical treatment process that suspends particles from liquid by forcing fluid through a porous substance such as paper, cloth, fine clay, sand, or charcoal, entrapping suspended particles on or within the filter medium.

Final Cover   The cover material applied upon closure of a landfill that is permanently exposed at the surface.

Finite Tolerance   The maximum amount of pesticide that can legally remain on a food or feed crop at harvest after the pesticide has been directly applied to the crop.

Fire   Rapid oxidation of a fuel; see COMBUSTION.

Fire Department Safety Officer   A member of a fire agency assigned and authorized by the fire chief to perform the duties and responsibilities defined in the adopted safety standard for the organization.

Fire Point   The lowest temperature of a liquid at which vapors are evolved fast enough to support continuous combustion; related to flash point.

Fireproof   A misnomer applied to building materials or structural components of a building.

Firescope (ICS)   Firefighting Resources of Southern California Organized for Potential Emergencies.

Fire Wall   A wall of earth built around an oil tank to hold the oil if the tank breaks or burns; a berm.

First Responder(s) (EMS)   Initial unit dispatched to the scene of a medical emergency to provide patient care.

First Responder, Awareness Level   Individuals who are likely to ness or discover a hazardous substance release and who have been trained to initiate an emergency response sequence by notifying the proper authorities of the release. They take no further action beyond noticying the authorities of the release. (CFR 29, 1910, 120)

First Responder, Operations Level   Individuals who respond to releases or potential releases of hazardous substances as part of the initial response to the site for the purpose of protecting nearby persons, property, or the environment from the eflects of the release. They are trained to respond in a defensive fashion without actually trying to stop the release. Their function is to contain the release from a safe distance, keep it from spreading, and prevent exposures (CFR 29, 1910.120).

Fissile   Material capable of undergoing fission.

Fissile Material   Any material fissionable by neutrons of all energies including and especially thermal (slow) neutrons as well as fast neutrons (uranium 235 and plutonium 239); term has a more restricted meaning than fissionable.

Fission   A process in which large radionuclides break into smaller pieces and release radiation in the form of particles or energy; the splitting of an atomic nucleus into two parts accompanied by the release of a large amount of radioactivity and heat.

Fission Products   The fragments (nuclei) formed by the fission of heavy elements, plus the nuclides formed by the fission fragments' radioactive decay.

Fissionable   A nucleus that undergoes fission under the influence of neutrons, even of very slow neutrons.

Fissionable Materials   Commonly used as a synonym for fissile material. Meaning also has been extended to include material that can be fissioned by fast neutrons only, such as uranium 238. Used in reactor operations to mean fuel.

Fittings   The small pipes and valves used to make up a system of piping.

Fixed Liquid Level Gauge   A gauge using a relatively small positive shutoff valve designed to indicate when the liquid level in a container being filled reaches the point of communication with the interior of the container.

Fixed Maximum Liquid Level Gauge   A fixed liquid level gauge that indicates the liquid level at which the container is filled to its maximum permitted filling density.

Fixed Signal   A signal of fixed location indicating a condition affecting movement of train or engine.

Flag Station   A station at which trains stop only when signalled.

Flame Impingement   The points where flames contact the surface of a container.

Flame-Spread   Speed at which a flame will cross the surface of a material; influenced by the physical form of the fuel, air supply, fuel's moisture content, specific gravity, size, and form, the rate and period of heating, and the nature of the heat source. A higher flame-spread critically affects the severity of the fire in a given period of time.

Flammability, Flammable   The capacity of ignition of a substance. Generally, the more flammable a substance, the more likely the spread of fire. In building materials, used in a general sense. In Class B liquids, applies to substances with flash points below 140°F. Commodity that can be easily ignited.

Flammable Gas   Any gas that will burn.

Flammable Liquid   Any liquid with flash point below 100°F (37.7°C).

Flammable Material   A substance capable of being easily ignited and of burning rapidly.

Flammable Range   Range of concentrations of a flammable vapor in the air. Lower flammable limit marks point at which vapors are too lean to burn; upper flammable limit marks the point at which the vapors are too rich to burn.

Flammable Solid   Any material, other than an explosive, liable to cause fires through friction or retained heat from manufacturing or processing, or that can be ignited readily; when ignited burns so vigorously and persistently as to create a serious transportation hazard.

Flange   A rib or rim for strength or guidance; a projecting edge on the circumference of a RR wheel to keep it on the rail.

Flash Point   Temperature at which a liquid gives off flammable vapors sufficient to form an ignitable mixture near the surface of the liquid; combustion is not continuous at the flash point.

Flashing Liquid-tight rail on top of a tank that contains water and spillage and directs it to suitable drains.

Flashing Drain Tubing that drains water and spillage from flashing to the ground.

Flat Car Open car without sides, ends or top, used primarily for hauling lumber, stone, and heavy machinery.

Flexible Connector A short (not exceeding 36 inches overall length) component of a piping system fabricated of flexible material such as hose, and equipped with suitable connections on both ends; used where there is a possibility of greater relative movement between points connected than is acceptable for rigid pipe.

Float Bridge A bridge connecting car floats with rail landings.

Float Gauge A gauge indicating the liquid level in a container constructed with a float inside the container resting on the liquid surface; transmits its position through suitable leverage to a pointer and dial outside the container.

Floatage The transfer of RR cars across water.

Floating Load (RR) A load in which the lading is prepared as a unit with space between unit and ends of car and end blocking omitted. Lengthwise movement of the lading over floor of the car permits the dissipating of impact shocks.

Flocculation A hazardous waste physical treatment method by which suspended particles are assembled into larger, more settleable particles after the waste is mixed with chemicals. This technique enhances the sedimentation process.

Flood Plain Lowland bordering a river which is usually dry but subject to flooding when the stream overflows.

Flora Plants in general; plant life as distinguished from animal life (fauna). Usually used in reference to all the plant life inhabiting or characteristic of a particular region or locality.

Flotation A hazardous waste physical treatment process by which particles are separated from liquid by introducing fine gas bubbles into the liquid, which attach to the particles and rise to the surface; particles are then collected by skimming mechanisms.

Flow Path The direction in which ground water is moving.

Flowable (1) Very finely ground solid material suspended in a liquid; usually contains a high concentration of the active ingredient and must be mixed with water when applied.

80

Fluid Resistant Clothing   Clothing designed and constructed to provide a barrier against accidental contact with body fluids.

Fly Ash   Fine particles of ash of a solid fuel that are either carried out of the flue with the waste gases produced during combustion or recovered from the waste gases.

Flycrew (ICS)   A handcrew of predetermined size transported to an incident via helicopter.

Foaming Agent   A material that causes a pesticide mixture to form a thick foam; used to reduce drift.

Fogger   An aerosol generator; pesticide equipment that breaks some pesticides into very fine droplets and blows or drifts the "fog" onto the target areas.

Foliar Application   Spraying a pesticide onto the stems, leaves, needles, and blades of grasses, plants, shrubs, or trees.

Food and Drug Administration (FDA)   Performs, directs, and coordinates detection and control activities that protect consumers against adulterated, misbranded or falsely advertised foods, drugs, medical devices, and hazardous products.

Food Chain   The dependency of one type of life on another, each in turn eating or absorbing the next organism in the chain.

Food-Chain Crops   Those crops grown for human consumption and pasture and other crops grown for feed for animals, whose products or by-products will be used for human consumption.

Force   A vector quantity of energy that tends to produce an acceleration in the direction of its application.

Fork Pockets   Transverse structural apertures in the base of the container that permit entry of fork lift devices.

Formations   Geologic strata that underlie the ground surface.

Formulation   The pesticide product; may contain one or more active ingredients, the carrier (if needed), and other additives (if needed) to make it ready for sale in a form that is safe and easy to store, dilute, and apply.

Formulation Plant   Plant designed to combine the technical pesticide with a solvent or diluent in order to prepare the material for commercial use.

Fractional Detonation   Partial detonation of a high explosive, resulting in the scattering of device parts and undetonated explosives over a wide area.

Fractional Distillation   Separation of a mixture of liquids having different boiling points, a primary process in the refining of crude oils.

Fracture   A break in the geological formation, e.g., a shear or fault.

Framing Lumber   Material used in construction and framing of houses. Usually pressure-treated with pentachlorophenol, a wood preservative.

Freeboard   (1) Vertical distance between top of a tank sidewall or lowest elevation of a surface impoundment dike or berm, and elevation of the highest surface of the waste contained in the tank or impoundment. (2) Part of a floating boom designed to prevent waves from washing oil over the top.

Free Liquids   Liquids that readily separate from the solid portion of a waste under ambient temperature and pressure.

Free Product   Fuel product accumulated on top of the ground water that is recoverable by well withdrawal methods. Free product is often mobile.

Free Time   The time allowed by the carriers for the loading and unloading of freight after which demurrage or detention charges will accrue.

Freight Agent   RR representative who prices services performed based on approved tariffs.

Freight Bill   Statement given to customer for charges of transportation; information is taken from waybill.

Freight Charge   Assessment for transporting freight.

Freight Claim   Demand upon a carrier for payment of overcharge, loss, or damage sustained by shipper or consignee.

Freight Classification   see CLASSIFICATION, and UNIFORM FREIGHT CLASSIFICATION.

Freight Forwarder   An individual or organization engaged in business of shipping and distributing less than a carload of freight.

Freight House   The station facility of a transportation line for receiving and delivering freight.

Frog   (1) A track structure used at the intersection of two running rails to provide support for wheels and passageways for their flanges, thus permitting wheels on either rail to cross to the other. (2) An implement for rerailing car wheels.

Front-End Loader   A refuse truck that has power-driven loading equipment at front of the vehicle.

82

Fuel   Any combustible substance that is burned to produce useful heat energy.

Fuel Element   A rod, tube, plate, or other mechanical shape or form into which nuclear fuel is fabricated for use in a reactor.

Fuel Oils   Refined petroleum products having specific gravities; in the range of 0.85 to 0.98 and flash points greater than 55 °C; includes furnace, auto diesel, and stove fuels, plant or industrial heating fuels, and various bunker fuels.

Fuel Tender (ICS)   Vehicle capable of supplying fuel to ground or airborne equipment.

Fuel Value   The amount of potential energy to be released by a fuel in the combustion process; expressed in terms of BTUs per pound of fuel. Examples:

Substance	BTU/pound
Wood Shavings	8,248
Wool Rags	8,876
Cotton Rags	7,165
Asphalt	17,158
LPG	18,000

Full Protective Clothing   Clothing that prevents gases, vapors, liquids, and solids from contacting the skin; includes helmet, self-contained breathing apparatus (SCBA), coat and pants customarily worn by firefighters, rubber boots, gloves, bands around legs, arms and waist, and face mask, as well as covering for neck, ears, and other parts of head not protected by the helmet, breathing apparatus, or face mask.

Full Service Application   Brake application resulting from a reduction in brake pipe pressure at a service rate until maximum brake cylinder pressure is developed.

Full Trailer   A truck trailer constructed so that all its own weight and that of its load rests upon its own wheels. A semitrailer equipped with a dolly is considered a full trailer.

Fully Encapsulated Suits   Chemical protective suit that is designed to offer full body protection, including self-contained breathing apparatus (SCBA), is gas-tight, and meets the design criteria outlined in NFPA Standard #1991.

Fulminate   In medicine, to develop suddenly and severely, as in a disease.

Fume   A smoke, vapor, or gas.

Fumigant   A pesticide that enters the pest in the form of a gas and kills; may be a liquid that becomes a gas when applied.

83

Fumigation   Use of chemicals in form of gas to destroy noxious insects, nematodes, and unwanted plants.

Fungi (Fungus)   Groups of microorganisms that cause rots, molds, and plant diseases; lack chlorophyll; grow from seed-like spores and produce tiny thread-like growths. Some fungi are pathogenic, they can attack and destroy nonliving things. Examples: molds, mushrooms, and yeasts.

Fungicide   Pesticide that controls or inhibits fungus growth.

Fusee   Red flare used for flagging purposes.

Fusible Plugs   A safety relief device in form of a plug of a low-melting metal. Plugs close the safety relief device channel under normal conditions and are intended to yield at a set temperature to permit the escape of gas.

Fusion   (1) Transition of a material from solid to a liquid (melting). (2) A process in which small nuclei combine into larger nuclei and release radiation in the form of particles or energy.

G

G   Gram; 1/1000 of a kilogram.

Gauge of Track   The distance between the heads of the rails, measured at a point 5/8-inch below top of the rails; standard gauge is 4 feet 8-1/2 inches.

Gauging Nipple   A small section of pipe in the top of a tank through which a tank may be gauged.

Gallon   U. S. Standard unit of measure; One gallon = 0.833 Imperial gallons = 231 cubic inches = 3.785 liters.

GAMA   Gas Appliance Manufacturers Association.

Gamma   A type of electromagnetic radiation; a form of ionizing radiation.

Gamma Rays   High-energy, short-wavelength electromagnetic forms, comprised of photons or fine packets of energy, which travel in straight paths at the speed of light; very penetrating but do not make material radioactive. Best shielded by dense materials such as lead or depleted uranium.

Gamma Ray Irradiation   Experimental hazardous waste chemical treatment method that disinfects waste by utilizing gamma radiation to destroy pathogens (disease-causing microorganisms).

Gandy Dancer (slang)   RR track laborer.

Gantry Crane   A stilted traveling crane supported on a bridge or trestle. Trestle belts are constructed on wheels so the whole structure travels on a track laid on the ground or floor.

Garbage   Decayable animal and vegetable wastes resulting from handling, preparation, cooking, and consumption of food. Also, trash or solid waste; things discarded regardless of reusability or recyclability.

Gas   In the widest sense applied to all aeriform bodies, the most minute particles of which exhibit the tendency to fly apart from each other in all directions. Normally these gases are found in that state at ordinary temperature and pressure. They can only be liquefied or solidified by artificial means: either through high pressure or extremely low temperatures.

Gas-Air Mixer   A device or system of piping and controls that mixes LP-Gas vapor with air to produce a mixed gas of a lower heating value than the LP-Gas. The mixture may replace another fuel gas completely or may be mixed to produce similar characteristics and then mixed with the basic fuel gas.

Gas Mask   Type of respirator covering entire face, providing eye protection as well as protection for the nose and mouth; effective against air-containing sprays, dusts, or gases.

Gaseous Diffusion (Plant)   A method of isotropic separation based on the fact that gas atoms or molecules with different masses will diffuse through a porous barrier at different rates; requires large plants and enormous amounts of electric power.

Gasolines   A mixture of volatile, flammable liquid hydrocarbons used primarily for internal combustion engines; flash point of approximately $-40\,^{\circ}$C.

Gas Processing Plant   A facility designed (1) to achieve the recovery of natural gas liquids from the stream of natural gas which may or may not have been processed through lease separators and field facilities, and (2) to control the quality of the natural gas to be marketed.

Gastrointestinal Syndrome   Illness caused by acute exposure to ionizing radiation, resulting in damage to the gastrointestinal system.

Gateway   Point at which freight moving from one territory to another is interchanged between railroads.

GATT   General Agreement on Tariffs and Trade.

GC   Gas Chromatography.

GC/FID Gas chromatography/flame ionization detector.

GC/MS Gas chromatography/mass spectrometry.

GC/PID Gas chromatography/photoionization detector.

Geiger-Muller Counter (Tube) A radiation detection and measuring instrument, consisting of a gas-filled tube containing electrodes between which there is an electrical voltage but no current flowing. When ionizing radiation passes through, a short intense pulse of current passes from the negative to the positive electrode and is measured. Number of pulses per second measures the intensity of radiation. Also known as a Geiger Counter.

Gelling Agents Chemicals that increase the viscosity of oil, and in theory, can be applied to an oil slick to reduce its rate of spread over the water surface.

General Emergency (Nuclear Reactors) An event involving actual or imminent substantial core degradation (or melting) with potential for loss of containment integrity and with subsequent release of significant radioactivity to the environment. Off-site protective actions may be necessary.

General Service Car RR box, gondola, or flat car not designed for a specific commodity or shipper, without special equipment.

General Staff (ICS) A group of incident management personnel comprised of but not limited to the incident commander, suppression and rescue section chief, planning section chief, and logistics section chief.

Generator Person responsible for the creation of hazardous waste by nature of ownership, management, or control.

Genetic Effects of Radiation Radiation effects that can be transferred from parent to offspring; any radiation-caused changes in the genetic material of sex cells.

Geometry (1) The spatial configuration, pattern or relationship of components in an experiment or apparatus, (2) In reactor technology, refers to shape and size of fuel elements, moderator, reflector, and their location with respect to each other. (3) In nuclear physics, the arrangement of source and detecting equipment. (4) In counting and scanning, the percentage of radiation leaving a sample that reaches the sensitive volume of a counter.

Glad Hands End of an air hose.

Glasphalt Trade name for a highway paving material wherein recovered ground glass replaces some gravel normally used in asphalt.

Glass A material made from fusion of sand, soda ash, and other ingredients. Common glass is impermeable, transparent, sanitary and odorless.

Glove Box   A sealed enclosure in which all handling of items inside the box is carried out through long rubber gloves sealed to ports in the walls of the enclosure. Operator places hands and forearms in the gloves from the room side of the box to be physically separated from the gloved box environment, but is able to manipulate items inside the box with relative freedom while viewing the operation through a window.

Glycols   Any of a class of organic compounds belonging to the alcohol family but having two hydroxyl groups.

Goggles   Eye protection device.

Gondola Car   Freight car with sides and ends, but without a top covering.

Gooseneck   On a drop-frame trailer, the upper level at the front trailer together with the structure connecting it to the lower level. Also used on container chassis to reduce overall height of vehicle.

GOST   State Committee for Product Quality Central and Standards (former USSR).

Governor's Authorized Representative (Federal definition)   The person named by a governor in a federal/state agreement to execute, on behalf of a state, all necessary documents for disaster assistance. This person is designated after an emergency or major emergency disaster declaration is proclaimed by the president.

GPA   Gallons per acre.

GPM   Gallons per minute.

Grab Iron   Steel bar attached to cars and engines as a hand hold.

Grab Sample   A soil sample obtained without using a coring device. Frequently used during tank removals.

Gradient   (1) Pressure drop. (2) (Brake pipe) Difference in the brake pipe pressure between the front and rear of the train; direct result of leakage or line obstruction.

Grain Door   A partition placed across the door of a boxcar to prevent loss of grain by leaking.

Gram   The basic unit of weight in the metric system; equal to 1/1000 of a kilogram; approximately 28.5 grams equal one ounce.

Granular Pesticide   An active ingredient mixed with or coating small pellets or sand-like material; often used to control soil pests.

87

Granules   Dry, coarse particles of some porous material (clay, corn-cobs, walnut shells) into which a pesticide is absorbed.

Gravel Pack Envelope   Properly graded material used to backfill the annular space surrounding slotted casing.

Gravity (Specific)   Density expressed as the ratio of the weight of a volume of substance to the weight of an equal volume of a standard substance. In the case of liquids and solids, the standard is water; with natural gas or other gaseous materials, the standard is air.

Gray (Gy)   The SI unit for measuring absorbed doses of ionizing radiation. 1 Gy = 1 joule/kg = 100 rads.

Gross Ton   2,240 pounds; also called long ton.

Gross Ton-Mile   Movement of a ton of transportation equipment and contents a distance of one mile.

Gross Weight   The weight of a RR car and its contents.

Grounding   Method whereby activities that will generate static electricity will be grounded to prevent static electricity discharge.

Ground Support Unit Leader (ICS)   Person responsible for establishing and supervising staging area(s); controlling and dispersing manpower, equipment, and apparatus; providing first aid medical care needs of incident personnel; providing for fuel, maintenance, and repair of equipment and apparatus.

Groundwater   Water present below the soil surface and occupying voids in the porous subsoil; specifically, the porous layer that is completely saturated with water. Upper boundary is referred to as the water table.

Groundwater Plume   A body of contaminated groundwater originating from a specific source and influenced by the local flow pattern, density, and concentration of contaminant, and the character of the aquifer.

Groundwater Protection Standard   The level of contamination that triggers the need for corrective action measures.

Groundwater Quality   The specific chemical, physical, and biological properties of groundwater in a specific area. State and local standards determine its suitability as a drinking water supply.

Group   Organization level within the incident command system having responsibility for operations within a specific functional area, i.e., salvage, ventilation, hazmat [National Incident Management System (NIIMS)].

Growth Regulator   A pesticide that increases, decreases, or in some way changes the normal growth or reproduction of a plant or insect.

Guidelines   Informal, instructional state or federal agency directives explaining program regulations or policies not addressed clearly by law. Guidelines do not have the force of law.

H&SC   California Health and Safety Code.

Habitat   The sum total of the environmental conditions of a specified place occupied by an organism, a population, or a community.

Hack (slang)   Caboose.

Half-Life   A measure of the decay rate of an isotrope to another nuclear form; measured half-lives vary from millionths of a second to billions of years.

Halons   Fire suppressing gases composed of straight chain carbon atoms with a variety of halogen atoms attached.

Halogens   A chemical family that includes the gases fluorine, chlorine, bromine, and iodine.

Hammermill   A type of crusher used to break up waste materials into smaller particles; operates by using rotating and flailing heavy hammers.

Hand Brake   Apparatus utilized to manually apply brakes on a car or locomotive.

Hand Crew (ICS)   Predetermined individuals supervised, organized, and trained principally for clearing brush as a fire suppression measure. In the context of hazardous materials, may be used to identify personnel used to perform manual tasks associated with hazardous materials operations.

Hard Water   Water containing soluble salts of calcium, magnesium, and sometimes iron.

Harm   An injury or damage.

Hatch   An opening into a tank, usually through the top deck.

Hatch Plan   Schematic drawing of location of all cargo on a ship; also called a stowage plan.

Hay Tank   An enclosure filled with hay-like material used to filter oil out of water.

Hazard   The chance that injury or harm will occur to persons, plants, animals, or property. In pesticides, the risk of danger resulting from a combination of toxicity and exposure.

Hazard Area   A geographically identifiable area in which a specific hazard exists and presents a threat to life and property.

Hazard Assessment   A process used to qualitatively or quantitatively assess risk factors to determine incident operations.

Hazard Class   The eight classes of hazardous materials as categorized and defined by Department of Transportation in 49 CFR.

Hazardous Air Pollutant   An airborne pollutant that may cause or contribute to an increase in mortality or serious illness.

Hazardous Chemical   An explosive, flammable, poisonous, corrosive, reactive, or radioactive chemical requiring special care in handling because of hazards it poses to public health and the environment.

Hazardous Class   A group of materials designated by the DOT sharing a common major hazardous property, e.g., radioactivity, flammability.

Hazardous Material   A substance in a quantity or form posing an unreasonable risk to health, safety, and/or property when transported in commerce; a substance that by its nature, containment, and reactivity has the capability of inflicting harm during an accidental occurrence; characterized as being toxic, corrosive, flammable, reactive, an irritant, or a strong sensitizer and thereby poses a threat to health and the environment when improperly managed.

Hazardous Materials Categories:

COMBUSTIBLE LIQUID—Any liquid having a flash point above 100°F, and below 200°F, as determined by tests listed in Code of Federal Regulations 49, Sec. 173.115.

CONSUMER COMMODITY—A material packaged or distributed in a form intended and suitable for sale through retail sales agencies for use or consumption by individuals for purposes of personal care or household use; includes drugs and medicines.

CORROSIVE MATERIAL—Any liquid or solid, including powders, that cause visible destruction of human skin tissue; a liquid that has a severe corrosion rate on steel or aluminum.

ETIOLOGICAL AGENT—A viable microorganism or its toxin that causes or may cause human disease.

EXPLOSIVE—Any chemical compound, mixture, or device, the primary or common purpose of which is to function by explosion, with substantially instantaneous release of gas and heat.

FLAMMABLE GAS—Any gas which, in a mixture of 13% or less by volume with air, is flammable at atmospheric pressure, or its flammable range with air at atmospheric pressure is wider than 12% (by volume) regardless of a lower flammability limit.

FLAMMABLE LIQUID—Any liquid having a flash point below 100°F, as determined by tests listed in Code of Federal Regulations 49, Sec. 173.115(d).

FLAMMABLE SOLID—Any solid material, other than an explosive, that is liable to cause fires through friction or retained heat from manufacturing or processing, or which can be ignited readily and when ignited burns so vigorously and persistently as to create a serious transportation hazard.

IRRITATING MATERIAL—A liquid or solid substance that upon contact with fire or when exposed to air gives off dangerous or intensely irritating fumes, but not including any Class A poisonous material.

NONFLAMMABLE GAS—Any compressed gas other than a flammable gas.

ORGANIC PEROXIDE—An organic compound that may be considered a derivative of hydrogen peroxide where one or more of the hydrogen atoms have been replaced by organic radicals; readily releases oxygen to stimulate the combustion of other materials.

OXIDIZER—A substance that yields oxygen readily to stimulate the combustion of other material.

POISON A—A poison gas; extremely dangerous gases or liquids of such nature that a very small amount of the gas or vapor of the liquid mixed with air is dangerous or lethal to life.

POISON B—Liquids or solids, including gases, semi-solids, and powders other than CLASS A or IRRITATING MATERIALS, that are known to be so toxic to man as to afford a hazard to health.

RADIOACTIVE MATERIAL—Also known as Radiological Material, any material or combination of materials that spontaneously emits ionizing radiation, and having a specific gravity greater than 0.002 microcuries per gram.

Hazardous Materials Emergency The release or threatened release of a hazardous material that may impact the public health, safety, and/or the environment.

Hazardous Materials Incident Contingency Plan (HMICP) Established pursuant to Section 8574.7 of the California Government Code, fulfills the requirement for a state toxic disaster plan, and provides for an integrated and effective state procedure to respond to the occurrence of toxic disasters within the state.

91

Hazardous Materials Ordinance The ordinance adopted by the administering agency to regulate the construction, monitoring, and reporting at hazardous materials facilities.

Hazardous Materials Response Team (HMRT) An organized group of employees, designated by the employer, who are expected to perform work to handle and control actual or potential leaks or spills of hazardous substances requiring possible close approach to the substance. A hazmat team may be a separate component of a fire brigade or a fire department. Members must be hazardous materials specialists. (CFR 29 Section 1910.120).

Hazardous Materials Response Team Levels:

SPECIALIST LEVEL—Shall consist of an organized group of employees, designated by the employer in compliance with SARA Title III, Section 1920.120(q)(6), and trained to perform work in hazardous materials incident handling at the specialist level in accordance to NFPA Standard 472, Chapter 4. The team shall include sufficient personnel to ensure:

1. The provision of at least these positions in accordance to FIRE-SCOPE ICS-HM-120, trained to Specialist Level:
- Group Supervisor
- Entry Leader
- Safety Officer
- Site Access Control Officer
- Decontamination Leader

2. The provision of at least these positions, in accordance to SARA Title III, appropriately trained for entry with encapsulating vapor protective clothing:
- Entry Team—2
- Back-Up Team—2

The employer should be reminded that total appropriate staffing of the specialist team should also include provisions of other positions such as stenographer, decontamination personnel, and additional entry personnel.

SPECIALTY—Shall consist of an organized group of employees, designated by the employer in compliance with SARA Title III, Section 1910.120(q)(6), and trained in the hazards of specific hazardous substances, the provision of specialized technical advice, and specialized assistance, in compliance with SARA Title III, section 1910.120(q)(5). Examples of some highly specialized functions: Decontamination, off-loading, hot tapping, chlorine team response, etc. The team shall include sufficient personnel, either within their own team or in agreement with a Hazardous Materials Response Team—Specialist or a Hazardous Materials Response Team—Technician on scene, to ensure:

1. The provision of at least three positions in accordance to FIRE-SCOPE ICS-HM-120, appropriately trained:

- Group Supervisor
- Entry Leader
- Safety Officer
- Site Access Control Officer
- Decontamination Leader

2. The provision of at least two personnel in the entry team, appropriately protected associated with the hazard and with a back up team of two personnel similarly protected.

TECHNICIAN LEVEL — Shall consist of an organized group of employees, designated by the employer in compliance with SARA Title III, Section 1910.120(q)(6), and trained to perform work in hazardous materials incident handling at the technician level in accordance to NFPA 472, Chapter 3. The team shall include sufficient personnel to ensure:

1. The provision of at least these positions in accordance to FIRE-SCOPE ICS-HM-120, trained to the level as indicated:
- Group Supervisor Specialist
- Entry Leader Technician*
- Safety Officer Technician
- Site Access Control Officer Technician
- Decontamination Leader Technician

2. The provision of at least these positions in accordance to SARA Title III, appropriately trained to include at least entry with splash protective clothing:
- Entry Team — 2
- Back-Up Team — 2

*Entry Leader may on occasion function also as the group supervisor and must also be trained to specialist level.

The employers should be reminded that total appropriate staffing of the technician team should also include provisions of other positions such as stenographer, decontamination personnel, and additional entry personnel.

Hazardous Materials Safety Officer/Official A person at a hazardous materials incident responsible for assuring that all operations performed at a hazardous materials incident, by all members present, are done so with respect to the highest levels of safety; has full authority to alter, suspend, or terminate any activity that may be judged to be unsafe; reports to the hazardous materials group supervisor.

Hazardous Substance As used by California OSHA, encompasses every chemical regulated by both the DOT (hazardous materials) and EPA (hazardous waste), including emergency response (29 CFR 1910.120).

Hazardous Substances Account A California state fund derived from fees paid by persons who submit more than 500 pounds per year of hazardous or extremely hazardous waste to on- or off-site hazardous waste disposal facilities; funding source for the state Superfund program.

Hazardous Waste Waste materials or mixtures of waste that require special handling and disposal because of their potential to damage health and the environment.

Hazardous Waste Constituent Substance contained in waste causing that waste to be listed as hazardous.

Hazardous Waste Control Account An ongoing state fund, derived from fees paid by operators of on- and off-site hazardous waste disposal facilities, that is the basic funding source for the Department of Health Services' hazardous waste management program.

Hazardous Waste Control Act A California state law enacted in 1972; the first comprehensive law of this type in the United States, it established the state's hazardous waste management program within the Department of Health Services.

Hazardous Waste Element That portion of a county solid waste management plan that addresses hazardous waste management.

Hazardous Waste Facility A facility that handles, stores, treats, or disposes of a hazardous waste and that contains at least one hazardous waste area, exclusive of a facility using a biological process on the property of a producer treating oil, its products, and water and producing an effluent which is continuously discharged to navigable waters in compliance with a permit issued pursuant to Section 402 of the Federal Water Pollution Control Act (33 U.S.C. 1342). (Section 66096, Title 22, California Administrative Code; the Department of Health Services is proposing to revise this definition.)

Hazardous Waste Facility Permit Document issued by the Department of Health Services granting the authority to operate a transfer station or a facility that stores, treats, or disposes of a hazardous waste.

Hazardous Waste Generation The act or process of producing hazardous waste.

Hazardous Waste Landfill An environmentally sound disposal facility where hazardous waste can be placed without polluting the environment, not including a land treatment facility, surface impoundment, or injection well.

Hazardous Waste Leachate Any liquid that has percolated through or drained from hazardous waste placed in or on the ground.

Hazardous Waste Management Systematic control of the collection, source separation, storage, transportation, processing, treatment, recovery, and disposal of hazardous wastes.

Hazardous Waste Manifest, Uniform (EPA Usage) The shipping document, originated and signed by the waste generator or an authorized representative, that contains the information required; must accompany shipments of hazardous waste. (Title 40 CFR 262, Subpart B)

Hazardous Waste Site A location where hazardous wastes are stored, treated, incinerated, or otherwise disposed of.

HAZ-CAT Hazardous Material Categorization Test; a field analysis to determine the hazardous characteristics of an unknown material (FIRESCOPE HCS 120-1).

HAZ-MAT Hazardous materials.

HAZOP Hazard and Operability Study; a systematic technique for identifying hazards or operability problems throughout an entire facility. Examines each segment of a process and lists all possible deviations from normal operating conditions and how they might occur. The consequences of the process are assessed, and the means available to detect and correct the deviations are examined.

HAZWOPER Hazardous Waste Operator.

HCl Hydrochloric acid.

Head The front and rear closure of a tank shell.

Head End Beginning or forward portion of any train.

Head Man Brakeman responsible for work done in connection with the forward section of the train and, when in transit, is stationed in the locomotive; head pin (slang).

Head Space The air space at the top of a water or soil sample.

Head Wave An area of oil concentration that occurs behind and at some distance from containment booms; area is important to the positioning of mechanical recovery devices and is the region where droplet breakaway boom failure phenomenon is initiated when current flow exceeds critical velocity; see DROPLET BREAKAWAY.

Header Identifying portion (beginning) of any list or consist.

Health Facility (ICS) Any facility, place or building organized, maintained, and operated for the diagnosis, care, and treatment of human illness or injury, physical or mental, including convalescence, rehabilitation, and/or pre- and post-natal care, for one or more persons, to which patients are admitted for twenty-four hours or longer.

Health Hazard Any property of a material that can either directly or indirectly cause injury or incapacitation, either temporary or permanent, from exposure by contact, inhalation, or ingestion.

Health Hazard, Chemical Any chemical or chemical mixture whose

95

physical or chemical properties may cause acute or chronic health effect (29 CFR 1910).

Health Database A compilation of records and data relating to the health experience of a group of individuals, maintained in a manner such that it is retrievable for study and analysis over a period of time.

Heat A condition of matter caused by the rapid movement of its molecules. Energy has to be applied to the material in sufficient amounts to create the motion, and may be applied by mechanical or chemical means.

Heat (A Connection) To loosen a collar or other threaded connection by striking it with a hammer.

Heat of Fusion see LATENT HEAT.

Heat Measurement Temperature; may be calibrated on several scales, most commonly Fahrenheit and Centigrade; measurement of heat quantity is BTU.

Heat of Vaporization see LATENT HEAT.

Heat Transfer see RADIATION, CONVECTION, CONDUCTION.

Heater (Switch) A device for melting snow at switches by means of steam, an electric current, gas jets, or oil.

Heater Car An insulated box car equipped with heating apparatus for the protection of perishables.

Heating Tube A tube device installed inside a tank used to heat the tank contents; fire tube.

Heavy Equipment Transport (ICS) Any ground vehicle capable of transporting a bulldozer.

Heavy Metals High-density metallic elements generally toxic to plant and animal life in low concentrations (e.g., mercury, chromium, cadmium, arsenic, and lead).

Heavy Water Water containing significantly more than the natural proportion (one in 6,500) of heavy hydrogen (deuterium) atoms to ordinary hydrogen atoms; used as a moderator in some nuclear reactors as it slows neutrons effectively and has a low cross section for absorption of neutrons.

Hectare A metric area measurement equal to 10,000 square meters or approximately 2.47 acres.

Helibase, Helispot A location within an incident area where helicopters may be landed, serviced, and loaded.

Hemi- A prefix that indicates division into two parts or halves.

Hemispherical Head   Used on MC-331 high pressure tanks, head is in shape of a half-sphere.

HEPA Filter   *see* ABSOLUTE FILTER.

Hepatoxic   A substance that negatively affects the liver.

Hepta-   A prefix indicating the number seven.

Herbicide   Pesticide used to control plant life.

Hexa-   A prefix indicating the number six.

High Iron (slang)   Mainline or high speed track of a RR system.

High Rail   The outer or elevated rail of a curved track.

High Side Gondola Car   Gondola car with sides and ends over 36 inches.

High Expansion Foam   Detergent-base foam with low water content which expands in ratio of 1000 to 1.

Highball (slang)   Signal to proceed at maximum authorized speed.

HIMA   Health Industries Manufacturers Association.

HIT   Hazard Information Transmission program; provides a digital transmission of the CHEMTREC emergency chemical report to first responders at the scene of a hazardous materials incident. The report advises the responder on the hazards of the materials, the level of protective clothing required, mitigating action to take in the event of a spill, leak, or fire and first aid for victims. HIT is a free public service provided by the Chemical Manufacturers Association. Reports are sent in emergency situations only to organizations that have preregistered with HIT. Brochures and registration forms may be obtained by writing: Manager, CHEMTREC/CHEMNET, 2501 M Street, N.W., Washington, DC, 20037.

Hitch   A connecting device at rear of a vehicle used to pull a full trailer with provision for easy coupling.

HIV   Human Immunodeficiency Virus.

HMRT   *see* HAZARDOUS MATERIALS RESPONSE TEAM.

Hog (slang)   Locomotive.

Hog Head (slang)   Locomotive engineer.

Hogger (slang)   Locomotive engineer.

Hog Law (slang)   Federal statute providing that all train and engine crews must be relieved of duty after fourteen hours of continuous service (to be changed to twelve hours in future).

Hold Track   RR track where cars are held awaiting disposition.

Hole (slang)   An area of track enabling one train to pass another.

Home   A location where a RR car is on the tracks of its owner.

Home Car   A car on the tracks of its owner.

Home Junction (RR)   A junction with the home road.

Home Road   The owning road of a railroad car.

Home Route   Return route of a foreign empty RR car to the owning road.

Home Scrap   Scrap utilized within the plant where it originates.

Home Signal (RR)   A fixed signal at entrance to an interlocking to govern trains or engines entering and using that block.

Homo-   A prefix indicating a homologue of the compound to the name of which it is affixed and usually differing by having one more CH_2 in the formula, such as homosalicylic acid.

Hook (slang)   A crane used in wreck train service; also Big Hook, Wrecker.

Hopper   (1) An open-top RR car with one or more pockets opening on the underside of the car to permit quick unloading of bulk commodities. (2) Sloping panels at the bottom of a tank that direct dry bulk solids to the outlet piping.

Horsepower   A measure of power; one horsepower is equivalent to a force that will raise 33,000 pounds one foot in one minute.

Hose Tube   A housing used on tank and bulk commodity trailers for the storage of cargo-handling hoses; hose troughs.

Hostler   A RR fireman who operates light engines in designated enginehouse territory and works under the direction of the engine house foreman.

Hostler's Control   A simplified throttle provided to move the "B" unit of a diesel locomotive not equipped with a regular engineer's control.

Hot Box   An overheated journal caused by excessive friction between bearing and journal, lack of lubricant, or foreign matter.

Hot Box Detector   A wayside infrared sensing instrument for determining journal temperature.

Hot Cell   A heavily shielded enclosure in which radioactive materials can be handled remotely with manipulators and viewed through shielding windows to limit danger to operating personnel.

Hot Spot   Radioactive area.

Hot Tapping A method of welding on and cutting holes through liquid, compressed gas vessels, and piping for the purpose of relieving pressure and/or removing product.

Hot Zone An area immediately surrounding a hazardous materials incident that extends far enough to prevent adverse effects from hazardous materials releases to personnel outside the zone; also referred to as the exclusion zone, red zone or restricted zone in other documents (NFPA 472, 1–3).

HPLC High Performance Liquid Chromatography; used in organics analysis to separate chemical mixtures based on differential ionic absorption to various substrates.

H_2SO_4 Sulfuric acid.

Hump An incline in a RR yard over which cars are uncoupled and allowed to roll free into a classification yard.

HV Hydrocarbon Vapor.

Hy-Cube Car A box car of approximately 85-feet length and 10,000-cubic-foot capacity designed for hauling automobile body stampings and other low density freight.

Hydrapulper Trade name for a large mechanical device used primarily in paper industry to pulp waste paper or wood chips and separate foreign matter; suspends finely divided cellulose fibers in water.

Hydraulic Dispersion Shoreline cleanup technique utilizing a water stream at either high or low pressure to remove stranded oil; most suited for removal of oil from coarse sediments, rocks, and man-made structures.

Hydration Process in which particles go into a water solution and become surrounded by a sheath of water molecules.

Hydraulic Continuity A water bridge or connection between two or more geological formations.

Hydraulic Gradient The rate of inclination of groundwater surface that shows the direction of flow.

Hydrocarbon A chemical containing only carbon and hydrogen atoms. Crude oil is a mixture largely of hydrocarbons; methane (natural gas) is simplest, with each molecule containing one atom of carbon and four atoms of hydrogen.

Hydrogen The lightest element, No. 1 in the atomic series; has two natural isotopes of atomic weights 1 and 2; first is ordinary (or light) hydrogen, second is deuterium (heavy) hydrogen. A third isotope, tritium, atomic weight 3, is a radioactive form produced in reactors by bombarding lithium 6 with neutrons.

99

Hydrogen Bomb   Nuclear weapon deriving its energy largely from fusion.

Hydrogen Ion Concentration   Measure of acidity or alkalinity, expressed in terms of the pH of the solution; see pH.

Hydrogeologic Conditions   Those conditions dealing with the behavior of groundwater.

Hydrogeology   Scientific considerations relating to geological formations, soil surface water, and especially ground water.

Hydrology   Study of the properties, distribution, and flow of water on or within the earth.

Hydrolysis   Hazardous waste chemical treatment method wherein chemical compounds are decomposed by a reaction with water; agents such as alkaline solutions as well as high temperatures and pressures are often used to promote desired reaction.

Hydrophilic   "Water loving"; molecules that associate with water and are readily wet by water.

Hydrophobic   "Water hating"; molecules that are poorly soluble in water, are water repellant or not wet by water.

Hydrophobic Agent   Chemical having the ability to resist wetting by water; occasionally used in treatment of synthetic sorbents to decrease the amount of water absorbed, hence increasing the volume of oil they can absorb before becoming saturated.

Hydrophyte   A plant growing in water or soil too water-logged for most plants to survive.

Hydropunch   A water-sampling tool that is forced to a depth of about five to ten feet below the water table in order to retrieve a water sample through a one-way valve.

Hygroscopic   Ability of a substance to absorb moisture from the air.

Hyper-   A prefix meaning over, usually implying excess or exaggeration. Chemically, the same as super, indicating the highest of a series of compounds, such as hyperchloric acid. The prefix per- is now generally used for hyper- as in perchloric permanganic, etc.

Hypo-   A prefix meaning under, either in place or in degree (less or less than). A prefix in chemicals applied to the inorganic acids (as in hypochloric acid) and to their salts (as in potassium hypochlorite), to indicate a low valence state for the designated element.

Hypergolic   Any substance that spontaneously ignites upon contact with another. Many hypergolics are used as rocket fuels.

IAFC International Association of Fire Chiefs.

Iatrogenic "Caused by the doctor," a complication, injury, or disease state resulting from medical treatment.

IBN Institut Belge de Normalisation (Belgium).

ICAITI Central American Research Institute for Industry.

ICC Interstate Commerce Commission; independent federal government agency in the executive branch (not affiliated with DOT) charged with administering acts of Congress affecting rates and routes for transport of interstate commerce.

ICC Cylinder see CYLINDER.

Ice Bunker (Refrigerator car) Compartment where ice is placed.

Icing Charge Made for icing perishable freight.

ICS see INCIDENT COMMAND SYSTEM.

Identification Code The individual number assigned each hazardous waste generator, transporter, and treatment, storage or disposal facility by state or federal regulatory agencies.

Identification Lamps On the rear of a vehicle, a cluster of three red lamps mounted as close as practical to the top of the vehicle at the same height, one on the vertical centerline, one on each side of the vertical centerline, with lamp centers spaced not less than 6 inches nor more than twelve inches apart.

Idler Car An unloaded flat car used to protect overhanging loads or used between carrying cars loaded with long material.

IEMS Integrated Emergency Management System; developed by FEMA in recognition of the economies realized in planning for all hazards on a generic functional basis as opposed to developing independent structures and resources to deal with each type of hazard.

Ignitable Waste A liquid with a flash point lower than 60°C (140°F); a waste that is an oxidizer, ignitable compressed gas, or nonliquid which is liable to cause fires through friction, absorption of moisture, spontaneous chemical changes, or when ignited burns so vigorously and persistently as to create a hazard.

Ignition Temperature The temperature to which a substance must be heated in order to initiate self-sustained combustion (burning).

IGR Insect Growth Regulator.

101

Illegal Residue   A residue in excess of a pre-established and government-enforced safe level.

Immediately Dangerous to Life or Health (IDLH)   An atmospheric concentration of any toxic, corrosive, or asphyxiant substance that poses an immediate threat to life or would cause irreversible or delayed adverse health effects or would interfere with an individual's ability to escape from a dangerous atmosphere [29 CFR, Section 1910.120(a)(3)].

Imminent Hazard   An act or condition judged to present a danger to persons or property that is so urgent and severe that it requires immediate corrective or preventive action.

Immune   Exempt from or protected against; a state of not being affected by a disease or poison.

Immunization   The process of rendering a person immune from, or highly resistant to, a disease.

Impact Register   An appliance placed in a RR car with shipment that is both a time clock and a measuring device to record amount of shock the car received enroute.

Impermeable   Cannot be penetrated by gases or liquids; *see* SEMIPERMEABLE.

Impermeability   As applied to soil or subsoil, the degree to which fluids, particularly water, cannot penetrate in measureable quantities.

Import   To receive goods from a foreign country.

Improvised Booms   Booms constructed from readily available materials such as railroad ties, logs and telephone poles. Used as containment measures for hazardous materials spills.

-in   A suffix used in forming the names of fats, such as the glycerides or esters of glycerol; used as a suffix in the names of proteins and glucosides, such as amygdalin and albumin.

-ine   A suffix often used in forming the names of basic compounds such as amines, alkaloids and amino acids; examples: aniline, quinine, and glycine.

In-Bond Shipment   An import or export shipment that has not been cleared by federal customs officials.

In the Clear (RR)   When a car or train has passed over a switch and frog so far that another car or train can pass without collison.

In the Hole (slang)   On a siding.

In-Plant Waste   Waste generated in manufacturing processes; might be recovered through internal recycling or through a salvage dealer.

102

Inactive   Not involved in action. In pesticides, not reacting with anything.

Inactive Facility   The EPA designation for a treatment, storage, or disposal facility that has not accepted hazardous waste since November 19, 1980.

Inactive Portion   A portion of a hazardous waste management facility that has not operated since November 19, 1980, but that is not yet a closed portion (no longer accepts waste to that area).

Inbound Train   Train arriving at a yard or terminal.

Incentives   Measures providing benefits to communities above and beyond costs associated with hazardous waste management facilities. Also refers to certain measures (such as low interest loans, etc.) taken by government to stimulate development and implementation of improved hazardous waste management technologies.

Incident   An occurrence. Involving radioactive materials, an event resulting in the loss of control of radioactive materials; involves an immediate or likely hazard to life, health, or property.

Incident Action Plan (ICS)   A preplan containing general control objectives reflecting the overall incident strategy and specific suppression and rescue action plans for the next operational period.

Incident Command   A disciplined method of management established for the specific purpose of control and direction of resources and personnel.

Incident Command Post   see COMMAND POST.

Incident Command System (ICS)   An organized system of roles, responsibilities, and standard operating procedures used to manage and direct emergency operations. The incident command system in California is based on FIRESCOPE.

Incident Commander   The individual having the responsibility for total operations at a hazardous materials emergency.

Incineration   The application of high temperatures (800° to 3000°F) to break down organic wastes into more simple forms and to reduce the volume of waste needing disposal. Energy can be recovered from incineration heat.

Incineration Technologies   The processes by which waste volume is reduced by combustion in a controlled manner, their primary purpose being to thermally break down hazardous waste.

CEMENT KILNS—Used in the production of cement and in which organic wastes can be burned at temperatures ranging from 2600° to 3000°F.

LIQUID INJECTION SYSTEMS—Brick-lined chambers or beds of inert granular material into which solid and liquid wastes are injected for incineration.

ROTARY KILNS—Large, refractory cylinders capable of burning virtually any solid or liquid organic waste.

Incinerator   A plant designed to reduce waste volume by combustion. For pesticides, a special high-heat furnace that reduces contents to nontoxic ashes and vapors; used for disposing of highly toxic pesticides.

Incipient Fires   Fires that are in the beginning stages.

Incompatible   Not agreeable; not capable of being mixed or used together. When two or more pesticides are mixed and effectiveness of one or more is decreased; when two or more pesticides are mixed and application causes unintended injury to plants or animals.

Incompatible Waste   (1) A hazardous waste unsuitable for placement within a specific portion of a landfill because it may cause containment material to corrode or decay or, when combined with other wastes, might produce heat, pressure, fire, explosion, violent reaction, toxic dusts, mists, fumes, or gases. (2) Hazardous wastes that, if mixed, would become more hazardous than either waste would be individually.

Incompatibilities   A list of substances and conditions that can produce undesirable and potentially dangerous effects when stored, packaged, mixed, or handled together.

Incubation Period   The time from exposure to the disease until the first appearance of symptoms.

Independent Brake Valve (Air Brake)   Device used to operate the locomotive brakes independently of the train brakes.

Indication   Information conveyed by the aspect of a RR signal.

Indirect Disease Transmission   When a communicable disease is transmitted from one person to another without direct contact.

Individual Container   Cargo container, such as a box or drum, used to transport materials in small quantities.

Induce Vomiting   Make a person or animal throw up stomach contents.

Induced Radioactivity   Radioactivity created when substances are bombarded with neutrons, as from a nuclear explosion or in a reactor, or with charged particles produced by accelerators.

Industrial Hygienist   A qualified person responsible for recognition of hazards, identification of controls, calibration of equipment, interpretation of standards, collection of samples, and preparation of health and safety plans.

Industrial Plant (LP-Gas) A facility utilizing gas incidental to plant operations with LP-Gas storage of 2,000 gallons water capacity or more, and receiving gas in tank car, truck transport, or truck lots. Normally LP-Gas is used through piping systems in the plant, but may also be used to fill small containers such as for engine fuel on industrial (e.g., forklift) trucks. Because only plant employees have access to these filling facilities, are not considered distributing points; *see* DISTRIBUTION POINT.

Industrial Scrap Waste generated during a manufacturing operation; *see* IN-PLANT WASTE.

Industrial Waste Waste materials generally discarded from industrial operations or derived from manufacturing processes; may be liquid, sludge, or solid wastes and need not be hazardous.

Inert Any substance that will not react chemically; inactive.

Inert Ingredient Inactive ingredient; substance in a pesticide product that has no killing or controlling action.

Infection Growth of pathogenic organisms in the tissues of a host, with or without detectable signs of injury.

Infection Control Officer A member of a department assigned the specific responsibility for department infection control practices. This includes, but is not limited to, overseeing immunizations, post-exposure follow-up protocols. Fulfills the responsibilities for Designated Officer listed in the Ryan White Act.

Infection Control Program (ICP) A department's or agency's oral or written policy and implementation procedures relating to the control of infectious disease hazards where employees may be exposed to direct contact with body fluids (OSHA).

Infectious Capable of causing infection in a suitable host.

Infectious Disease An illness or disease resulting from the invasion of a host by a disease-producing organism such as bacteria, virus, fungi, or parasites.

Infectious Waste Waste containing pathogens; may consist of tissues, organs, body parts, blood, and body fluids removed during surgery.

Infiltration Flow of a fluid into a substance through pores or small openings; commonly used to denote flow of water into soil material.

Inflammable Liquids Liquids emitting vapors that become combustible at a certain temperature.

Ingestion   The taking in of toxic materials through the mouth. To eat, swallow, drink, or in some other way take into the digestive system.

Ingestion Pathway   Possible route by which radioactive material is introduced into the environment and subsequently ingested by living beings; principal exposure would be contaminated water or foods, especially near a nuclear facility.

Inhalation   The process of absorbing toxic materials by breathing in through the nose or mouth; taking air into the lungs; breathing in.

Inhibitor   Substance added to an unstable material that prevents its entrance into a violent reaction; a stabilizer.

Initial Attack (ICS)   Resources committed at the beginning of an incident.

Initial Carrier   Railroad on which a shipment originates.

Initial Point   Location at which a shipment originates.

Injection   Subsurface placement of a fluid or waste.

Injection Well   A well into which fluids are forced.

Injury   Disruption of the intended functional continuity of an animate or inanimate thing; may range from negligible to fatal.

Injury Mechanism   Process by which a material interacts to injure.

Inner Liner   Continuous layer of material placed inside a tank or other container to protect the construction materials of that container from the contents.

Inorganic   Substance that does not contain carbon.

Inorganic Compounds   Chemical compounds not containing carbon.

Inorganic Pesticides   Pesticides that do not contain carbon; examples: Bordeaux mixture, copper sulfate, sodium arsenite, sulfuric acid, and salt.

Insect Growth Regulator   A synthetic organic pesticide that mimics insect hormonal action so that the exposed insect cannot complete its normal development cycles and dies without becoming an adult.

Insecticide   Chemical used to control insect pests.

Insulated Rail Joint   A rail joint arresting the flow of electric current from rail to rail, as at the end of a track circuit, by means of nonconductors separating rail ends and other metal parts.

Integrated Control   Use of more than one method of pest control, including cultural practices, natural enemies, and selective pesticides.

Intensity   Amount of radiation passing through a unit area at a given time; measured in terms of roentgens by radiological survey instruments.

Inter-   Between.

Interchange   Exchange of cars between railroads at specified junction points.

Interchange Point   Location where RR cars are transferred from one road to another.

Interchange Track   Track on which freight is delivered by one railroad to another.

Interface Well   Vadose or unsaturated zone monitoring well, conventionally constructed with the screened or perforated interval located at a formational interface, usually permeable materials overlying impermeable or slightly permeable materials (i.e., in the case of tank backfill material contacting less permeable natural sediments).

Interim Authorization   Conditional permission from EPA that enables a state to operate its own hazardous waste management program.

Interim Status   A period of time when hazardous waste storage, treatment facilities, and transporters could continue to operate under a special set of regulations until the appropriate permit or license application is or was approved by DER.

Interline   Between one or more railroads.

Interline Freight   Freight moving from point of origin to destination over lines of two or more railroads.

Interline Waybill   Waybill covering movement of freight over two or more railroads.

Interlocking   An arrangement of RR signals and switches set in such a way that their movements must succeed each other in a predetermined order so that a clear indication cannot be given simultaneously on conflicting routes. Found at a crossing of two railroads, drawbridge, junction, or entering or leaving a terminal or yard.

Interlocking Limits   Tracks between the extreme opposing home signals of an interlocking.

Intermediate Carrier   A railroad over which a shipment moves but on which neither the point of origin nor destination is located.

Intermodel Container (ISO)   An article of transport equipment that meets the standards of the International Organization for Standardization (ISO) designed to facilitate and optimize the carriage of goods by one or more modes of transportation without intermediate handling of the contents and equipped with features permitting ready handling and transfer from one mode to another. Containers may be fully enclosed with one or more doors, open top, tank, refrigerated, open rack, gondola, flatrack, and other designs. Included in this definition are modules or arrays that can be coupled to form an intrinsic unit regardless of intention to move single or in multiplex configuration.

Internal Radiation   Radioactivity originating within the body. Usually deposited in the body by inhalation or ingestion of a radioactive contaminant.

Internal Valve   A primary shutoff valve for containers that has adequate means of actuation and is constructed in such a manner that its seat is inside the container; damage to parts exterior to the container or mating flange will not prevent effective seating of the valve.

International Air Transport Association (IATA)   An association of air carriers that develops guidelines for transportation of cargo.

International Civil Aviation Organization (ICAO)   Develops the principles and techniques of international air navigation and fosters the planning and development of international air transport so as to ensure safe and orderly growth.

Interstate Commerce Act   Act of Congress regulating practices, rates, and rules of transportation lines engaged in handling interstate traffic.

Interstate Commerce Commission   Regulating body of the United States government having jurisdiction over transportation matters.

Interstate Traffic   Traffic moving from a point in one state to a point in another state or between points in the same state, but passing within or through another state enroute.

Interval   The time between two occurrences; time between two pesticide applications; the period between the final pesticide application and harvest.

Intra-   Within.

Intra-Plant Switching   Moving of RR cars from one place to another within the yards of a plant or industry.

Intrastate Traffic   Traffic having origin, destination, and entire transportation within the same state.

108

Inverse Square Law   Mathematical relationship stating that radiation intensity is inversely proportional to the square of the distance from the source; as the distance doubles, the intensity decreases by one-fourth.

Invert Emulsifier   An adjuvant allowing water to remain suspended in oil rather than settling out; reverse of the usual emulsifier that allows suspension of oil in water.

Investigate   To systematically search or inquire into the particulars of an incident and collect the necessary evidence to seek criminal and/or civil prosecution.

Ion   A small particle with a net positive or negative electrical charge.

Ion-Pair   Produced when one or more electrons is separated from an atom. The electron is one-half of the ion pair, the nucleus the other.

Ionization   Process whereby neutral atoms or groups of atoms become electrically charged, either positively or negatively, by the loss or gain of electrons; process of producing ion-pairs.

Ionization Chamber   An instrument detecting ionizing radiation by measuring the electrical current that flows when radiation ionizes gas in a chamber, making the gas a conductor of the electricity.

Ionizing Radiation   Energy that interacts with matter by forming ion-pairs.

IPQ   Instituto Purtugues de Qualidade (Portugal).

IR   Infrared (a light wavelength).

Iron (slang)   RR switch.

Irradiation   Exposure to radiation, as in a nuclear reactor or fallout field.

Irritant   A material that has an anesthetic, irritating, noxious, toxic, or similar property and that can cause extreme annoyance or discomfort.

Irritating Materials   Liquids or solid substances that give off dangerous or intensely irritating fumes upon contact with fire or when exposed to air.

ISO   Insurance Service Organization; an organization that provides information on rating systems on fire service agencies to member-subscriber companies.

ISO   International Organization for Standardization; an international standards-writing body headquartered in Geneva, Switzerland, composed of national standards associations from some fifty-five countries. All member countries are given equal status and are entitled to one vote regardless of size or economic development. Technical work is carried on in committees.

109

Isolated Lenses   A body of ore or rock that is thick in the middle and thin at edges, like a double convex lens.

Isolated Source   A single, usually small, portable, and self-contained source of ionizing radiation, often located in an area where no other sources are present; examples: smoke detectors, radiopharmaceuticals, and radiography cameras.

Isolating the Scene   The preventing of persons and equipment from becoming exposed to a release or threatened release of a hazardous material by the establishment of site control zones.

Isolation Perimeter   Area around any hazardous materials incident within which only necessary personnel with full protective gear are allowed.

Isomer   One or more substances with the same composition but with different properties.

Isotonic   A solution having the same osmotic pressure as blood.

Isotope   One of two or more atoms with the same atomic number (the same chemical element) but with different atomic weights; usually have very nearly the same chemical properties but somewhat different physical properties.

Isotopic Enrichment   A process by which the relative abundances of the isotopes of a given element are altered, thus producing a form of the element that has been enriched in one particular isotope; example: enriching natural uranium in the uranium 235 isotope.

-ite   A suffix used in various ways for esters of the -ous acids (such as anylnitrite), and polyhydric alcohols (such as nannite), or explosives (like cordite, a smokeless powder).

-ium   A suffix used to form the names of metallic elements.

IV   Intravenous.

Jacket   Metal cover that protects the tank insulation.

Jackknife   Condition of truck tractor–semitrailer combination when their relative positions form an angle of ninety degrees (or less) about the trailer kingpin.

Jet Fuel   A kerosene or kerosene-based fuel used to power jet aircraft combustion engines.

Jib (slang)   Derrick or crane; boom.

Joint Agent   Person having the authority to transact business for two or more railroads.

Joint Bar   Splice bar; used in pairs, one on each side of the rail to fasten together the ends of rails; designed to fit the space between head and base closely and held in place by track bolts and suitable accessory equipment.

Jointly Liable   When two or more persons or companies share legal responsibility for negligence.

Joule   In physics, a unit of work or energy equivalent to one watt second.

Journal   End of the axle that moves in the bearings.

Journal Bearing   A combination of rollers and races or a block of metal in contact with the journal on which the load rests. In RR car construction the term when unqualified means a car axle journal bearing. *see* ROLLER BEARING.

Journal Box   Metal housing enclosing the journal of a car axle, the journal bearing, and wedge and holding the oil and lubricating device for lubricating the journal.

Judicial Review   A consideration by the courts concerning administrative agency decisions and actions.

Jumper   Flexible cable composed of one or more conductors of electric current used to connect electrically the controller circuits between cars or locomotives.

Junction Point   (1) Point at which a branch line RR track connects with a main line track. (2) Location at which two or more railroads interchange cars over connecting tracks.

Jurisdiction Specific Plan   A plan that details emergency activities, capabilities, responsibilities, and resources within an area, agency, facility, or political subdivision.

Jurisdictional Agency (ICS)   The agency having fire protection responsibility for a specific geographical area.

K

Keeper (RR slang)   Latch.

Kerosene   Flammable oil characterized by a relatively low viscosity

111

and flash point near 55 °C; lies between gasolines and fuel oils in terms of major physical properties.

Kg Kilogram.

Kick (RR slang) Act of pushing one or more cars at a speed sufficient to allow free forward movement into selected tracks when uncoupled and engine power reduced.

Kill a Well To overcome pressure in an oil well by use of mud or water so surface connections may be removed.

Kilo- Prefix meaning 1,000.

Kilogram Metric weight measurement; 1 kg = 1000 grams or approximately 2.2 pounds.

Kilton Energy The energy of a nuclear explosion equivalent to that of an explosion of 1000 tons of TNT.

Kinetic Energy Energy available due to motion.

Kingpin Attaching pin on a semitrailer that mates with pivots within the lower coupler of a truck tractor or converter dolly while coupling the two units together.

Knockout A kind of tank or filter used to separate oil and water.

Known Damage Damage discovered before or at the time of delivery of a shipment.

Knuckle (Coupler) Pin holding the knuckle in the jaws of the coupler; pivot pin.

Knuckle Thrower Device that throws the knuckle of a car coupler open when the uncoupling lever is operated.

L

Label Written or printed matter accompanying an article or substance to furnish identification; in pesticides, contains technical information.

Labeling Technical information about a substance in form of printed material provided by the manufacturer or its agent, including the label, flyers, handouts, leaflets, and brochures.

Labels (1) Four-inch square diamond markers required on individual shipping containers that are smaller than 640 cu. ft. (2) Diamond-shaped warning signs that are required to be placed on all containers of radioactive material; see PLACARD.

Labpack   Putting multiple small containers of chemicals with compatible chemical characteristics in a disposal drum with absorbent material.

Lacrimation   Tearing produced by eye irritation.

Ladder (RR)   Main track of a yard from which individual tracks lead; lead track.

Lading   Freight or cargo making up a shipment.

Lagoon   Shallow pond used for the temporary or permanent storage of liquid; sunlight, bacterial action, and oxygen interact to restore wastewater to a reasonable state of purity.

Land Burial   Disposal of wastes into land (landfills, surface impoundments, permanent waste piles, and underground injection wells); used for hazardous substances requiring permanent storage.

Land Disposal   Practice of disposing of liquid and solid hazardous wastes in earthen pits. In California, the State Water Resources Control Board licenses two classes of land disposal facilities that can accept hazardous wastes:

CLASS I—Sites that cannot overlie usable ground water, except in "extreme cases," and that may receive all classes of hazardous wastes except PCBs and radioactive wastes

CLASS II-1—Sites that may overlie or be adjacent to usable ground water, but must protect ground water by natural site characteristics or site modifications.

Land Disposal Methods of Dealing with Hazardous Waste Materials:

• landfill, surface impoundment, waste piles, deep well injection, land spreading, and co-burial with municipal garbage.

• treatment such as neutralization and evaporation ponds and land farming where residues are hazardous and are not removed for subsequent processing or disposal within one year

• storage such as waste piles and surface impoundments for longer than one year

Land Disposal Restrictions   A California state program administered by the Department of Health Services designed to progressively ban the land disposal of certain hazardous wastes.

Landfill   Facility for disposal of solid or hazardous waste involving burial in an excavated area or natural depression; sanitary landfill.

Landfill Cell   Compacted solid waste enclosed by soil or cover material within a landfill.

Land Treatment Facility   A location at which hazardous waste is applied or incorporated into the soil surface; such facilities are disposal facilities if the waste will remain after closure.

Larva   A young growing insect in the stage after hatching from the egg and before becoming a pupa; worm- or grub-like, usually not resembling the adult stage. Many insects cause most or all of their damage as larvae.

Larvacide   A pesticide used to kill the larvae of insects.

Latent Heat   The amount of heat absorbed or given off by a substance as it passes from a liquid to a gas (see HEAT OF VAPORIZATION), or from a solid to a liquid (see HEAT OF FUSION); measured in terms of the number of BTUs required to complete the phase.

Latent Period   A symptom-free period in acute radiation syndrome lasting from three to twenty-one days; follows prodromal period.

Lateral Motion (RR)   Motion crosswise of the track, that takes place in all car parts except the wheels and axles; results from the flexibility which must be provided in truck structure to permit safe negotiation of track curves.

Lawful Rate   A rate published in conformity with the provisions of the regulatory law and which does not violate any other provisions of such law.

LC_{50}   Concentration of an active ingredient in the air that, when inhaled, kills half of the test animals exposed to it; expression of a compound's toxicity when present in the air as a gas, vapor, dust, or mist; generally expressed in parts per million when a gas or vapor, and in micrograms per liter when a dust or mist; often used as the measure of acute inhalation toxicity. The lower the LC_{50} number value, the more poisonous the pesticide.

LD_{50}   Dosage or amount of an active ingredient, when taken by mouth or absorbed by the skin, that kills half of the test animals exposed; an expression used to measure Acute Oral or Acute Dermal Toxicity.

LDI   Leak Detection Investigation; part of the LDP to determine if contamination exists.

LD_{100}   The dose of an active ingredient taken by mouth or absorbed by the skin that is expected to cause death in 100% of the test animals so exposed.

LDP   Leak Detection Program.

Leachate A liquid that has percolated through solid waste or other material and carries decomposed waste, bacteria, and other noxious and potentially harmful materials which drain from landfills in solution or suspension and must be collected and treated so as not to contaminate water supplies. Groundwater contaminated by leachate is difficult to clean up, and it may take decades for the contaminants to be flushed out naturally.

Leaching Movement of a substance downward or out of the soil as the result of water movement.

Lead Agency The public organization having principal responsibility for approving a project or executing a plan.

Lead Track An extended track connecting either end of a yard with main track; ladder.

Leak The uncontrolled release of a hazardous material that could pose a threat to health, safety, and/or the environment.

Leak Control Compounds Substances used for plugging and patching leaks in nonpressure containers.

Leak Control Devices Tools and equipment used for plugging and patching leaks in nonpressure and some low pressure containers, pipes, and tanks.

Leakproof Bag A bag designed for disposal of potentially infectious substances; color coded and labeled in accordance with applicable laws.

Leaks and Spills Procedures Utilizes Hazardous Materials Storage and Transportation guidelines to provide instructions for dealing with air, water, and land spills and contamination by infectious agents. Spells out DOT firefighting recommendations and evacuation instructions; gives Hazardous Materials Table and Optional Hazardous Materials Table classification for the labeling, transportation, and storage of hazardous materials.

Legal Residue A remainder of a substance that is within a safe level according to the regulations.

LEL Lower Explosive Limit.

Less than Carload (RR) Quantity of freight less than that required for the application of a carload rate.

Less than Carload Rate (RR) Rate applicable to a less than carload shipment.

Lethal Causing or capable of causing death; deadly; fatal.

Lethal Concentration Amount of toxic substance in air that is likely to cause death if inhaled.

115

Lethal Dosage (1) Amount of a toxic substance that is likely to cause death when ingested. (2) Dose of ionizing radiation sufficient to cause death; medial lethal dose (MDL or LD_{50}) is amount required to kill within a specified period of time (usually 30 days) half of the organisms exposed; the LD50/30 for people is about 400–450 roentgens.

Leukemogen A substance that causes leukemia.

Leukocytes White blood cells.

Level of Protection In addition to positive pressure breathing apparatus, designations of types of personal protective equipment to be worn based on NFPA standards:

• LEVEL A—Vapor-protective suit for hazardous chemical emergencies

• LEVEL B—Liquid splash protective suit for hazardous chemical emergencies

• LEVEL C—Limited-use protective suit for hazardous chemical emergencies

Level One Incident Hazardous materials incident that can be correctly contained, extinguished, and/or abated utilizing equipment, supplies, and resources immediately available to first responders having jurisdiction, and whose qualifications are limited to and do not exceed the scope of training as explained in 29 CFR, Section 1910, or California Government Code (CGC), Chapter 1503, with reference to "First Responder, Operational Level."

Level Two Incident Hazardous materials incident that can only be identified, tested, sampled, contained, extinguished, and/or abated utilizing the resources of a Hazardous Materials Response Team, requires the use of specialized chemical protective clothing, and whose qualifications are explained in 29 CFR, Section 1910, or California Government Code (CFC), Chapter 1503, with reference to "Hazardous Materials Technician Level."

Level Three Incident A hazardous materials incident that is beyond the controlling capabilities of a Hazardous Materials Response Team (Technician or Specialist Level) whose qualifications are explained in 29 CFR, Section 1910, or California Government Code, Chapter 1503; and/or that requires the use of two or more Hazardous Materials Response Teams; and/or that must be additionally assisted by qualified specialty teams or individuals.

Liable Legally responsible.

Limited Mobility Population Persons requiring transportation during emergency movement operations.

116

Lithology The description of rock or soil in outcrop or hand sample on the basis of such characteristics as color, mineralogic composition, and grain size.

LOAEL Lowest Observable Adverse Effects Level; an estimate of dose or exposure threshold levels for toxic effects. It is determined from observations reported in a specific scientific study; *see* NOAEL.

Local Assessment Committee Review group created by a host or abutting community to analyze a proposed hazardous waste management facility. In some states such committees have the authority to negotiate with the facility proponent (on behalf of the community) regarding the conditions under which the hazardous waste management facility may be built.

Local Disaster Plan A plan developed and used by local government for extraordinary events.

Local Emergency The duly proclaimed existence of conditions of disaster or of extreme peril to the safety of persons and property within the territorial limits of a county, city and county, or city, caused by such conditions as air pollution, fire, flood, storm, epidemic, riot, earthquake, or other conditions, other than conditions resulting from a labor controversy, which conditions are or are likely to be beyond the control of the services, personnel, equipment, and facilities of that political subdivision and require the combined forces of political subdivisions to combat.

Local Emergency Planning Committee (LEPC) A committee appointed by the state emergency response commission, as required by Title III of SARA, to formulate a comprehensive emergency plan for its corresponding Office of Emergency Services mutual aid region.

Local Government A political subdivision within a state.

Local Oversight Program (LOP) The section of Los Angeles County Department of Public Works, Waste Management Division, in charge of overseeing cleanup of leaking underground storage tanks.

Local Veto Authority Within the context of hazardous waste management facility siting, refers to the ability of cities and counties to unilaterally reject proposed facilities by denying local land use approval.

Local Waybill A waybill covering movement of freight over a single railroad.

Localized Exposure Contact with a limited area, usually an external body surface.

Locomotive Engine on a train.

117

Logistics Section Chief (ICS)   Individual responsible at an emergency incident for management of units providing personnel, apparatus, equipment, facilities, and personal needs.

Long and Short Clause   Fourth Section of the Interstate Commerce Act; prohibits railroads from charging more for a shorter than for a longer haul over the same route, except by special permission of the ICC.

Long-Term Care   The post-closure monitoring and maintenance of a hazardous waste management facility in a manner that protects public health and the environment.

Long-Term Exposure   Exposure to radiation lasting more than four days.

Long Ton   2,240 pounds; also called a gross ton.

Look-Out   Caboose, cupola.

LOP   Local Oversight Program.

Lorry   Small four-wheel push car used in railroad construction and maintenance work for moving rails, ties, etc.

Low Concentrate Solution(s)   Mixture containing a small amount of an active ingredient in a highly refined oil; usually utilized as stock sprays and space sprays and for use in aerosol generators.

Low Population Zone   Area containing residents immediately surrounding an exclusion area when the number and density of that population is such that there is a reasonable probability that appropriate protective measures could be taken in their behalf in the event of a serious accident.

Low Rail   Inner rail of a curve that is maintained at grade while the opposite or outer rail is elevated.

Low-Side Gondola   Gondola RR car with sides and ends thirty-six inches high or less.

Low Specific Activity   Legal classification for radioactive shipments that allows the use of the radioactive LSA label in place of the radioactive I, II, or III label on sole-use vehicles; signified (1) natural radioactive materials such as ores, (2) materials with a high specific activity that are dispersed throughout a large volume of inactive material, or (3) nonradioactive items having a low concentration of surface contaminations. see ACTIVITY.

Lower Explosive Limit (LEL)   The lowest concentration of the material in air that can be detonated by spark, shock, fire, etc.

LPG   Liquefied Petroleum Gas.

LP-Gas Service Station Facility open to the public that consists of LP-Gas storage containers, piping and pertinent equipment, including pumps and dispensing devices, and any buildings in which LP-Gas is stored and dispensed into engine fuel containers of highway vehicles; *see* DISTRIBUTING POINT.

LSA Low Specific Activity.

Lubricating Oils Petroleum-based oils used to reduce friction and wear between solid surfaces such as moving machine parts and internal combustion engine components.

LUFT Leaking Underground Fuel Tank.

LUFT Manual A field manual to provide practical guidance to regulatory agencies.

Lunette Metal ring at end of tow bar for attachment of the tow bar to the towing vehicle.

LUST Leaky Underground Storage Tank.

M

Macroencapsulation The isolation of a waste by embedding it in or surrounding it with a material that acts as a barrier to water or air (e.g., clay and plastic liners).

Magnetic Gauge *see* FLOAT GAUGE.

Magnetic Separator Equipment used to remove magnetic materials from other materials, usually consisting of a belt, drum, or pulley and magnet.

Main Iron (RR) Main track.

Main Reservoir On RR air brakes, the tank on the engine for storing the main air supply, as opposed to the auxiliary reservoirs under each car.

Maintenance All actions taken to retain material in a serviceable condition to preserve serviceability; includes inspection, testing, servicing, classification as to serviceability, repair, rebuilding, and reclamation; routine, recurring work required to keep a facility in such condition to be continuously utilized in its original or designed capacity and efficiency.

Main Track A designated track upon which trains are operated by timetable, train order, or both, or the use of which is governed by block signals.

Make a Joint (Slang) Couple RR cars together.

Management (Hazardous Waste) A program for controlling the generation, storage, collection, transportation, treatment, use, conversion, or disposal of hazardous wastes; includes administrative, financial, legal, and planning activities as well as operational aspects of hazardous waste handling, disposal, and resource recovery systems.

Man-Machine System All of the personnel, equipment, and other resources capable of functioning during an emergency to return a community to normal conditions with the least amount of injury or damage.

Manhole Opening usually equipped with removable, lockable covers and large enough to admit a man into a tank trailer or dry bulk trailer.

Manifest A description of the contents of a shipment.

Manifest Illness Peak illness period of acute radiation syndrome lasting from twenty-one to forty-five days; follows the latent period.

Manifest Train Scheduled freight train.

Manifold Junction of a number of discharge pipelines to a common outlet.

Manipulators Mechanical devices used for safe handling of radioactive materials; frequently remotely operated from behind a protective shield.

Manpower and Training Component (EMS) Involves resources and arrangements for initial and continuing education and training of all personnel providing EMS; includes periodic proficiency testing, appropriate credentialing, and adequate career mobility and self-fulfillment opportunities.

Manual Block Signal System RR signal system wherein use of each block is controlled manually.

Manual Recovery Recovery of oil from contaminated areas by cleanup work force using buckets and shovels; extremely labor intensive.

Marked Capacity RR car load capacity as designated on that car.

Marker Front and rear signals of a train (flags or lamps).

Marking A sign, inscription, symbol, or visible impression on an article, such as a container.

Mass A measurement of the inertia of a body.

Mass Casualty (Disaster) (EMS) An emergency involving a large number of casualties and/or a response environment that has major barriers to the delivery of care; requires the response of multiple public agencies and delayed transport of the most seriously injured.

Mass Number   Total number of protons and neutrons in an atom.

MAST (EMS)   Military Assistance to Safety in Traffic; a program utilizing military helicopters and medical corpsmen to augment the existing EMS system.

MAST   (1) Medical Anti-Shock Trousers.

Master Gate   A large valve used to shut in an oil well.

Material   Substance. In pesticides, a formulation, chemical, active ingredient, or additive ingredient.

Materials Market   Combined commercial interests that buy substances. In the recycling industry, the market for products made from recycled materials determines the economic feasibility of recycling and resource recovery.

Material Safety Data Sheet (MSDS)   A document that contains information regarding the specific identity of hazardous chemicals, including information on health effects, first aid, chemical and physical properties, and emergency phone numbers.

Matter   That which has weight (mass) and occupies space; divided into three classes according to condition and state: solid, liquid, or gas. All changes matter exhibits are classified as physical or chemical; if composition is not altered in the process the phenomenon is said to be physical; if a change in composition takes place, it is referred to as chemical reaction. No destruction of matter is possible, only change to a different state. Often identified by its properties, such as density, conductivity, solubility, melting point, boiling point, hardness, color, etc.

Maximum Dosage   Largest amount of a pesticide that can be used safely without excess residues or damage occurring to whatever is being treated.

Maximum Rate   Highest RR rate that may be charged.

MCL   Maximum Contamination Level.

Meat Rack (RR refrigerator car)   Supports near the ceiling from which meat is suspended; beef rail.

Mechanical Agitation   Stirring, paddling, or swirling action of a device which keeps a pesticide and any additives thoroughly mixed in the spray tank.

Mechanical Refrigerator Car   A RR car equipped with a diesel-powered refrigerating unit under thermostatic control.

Mediation A voluntary negotiation process wherein a neutral mediator assists the parties in a dispute to reach a mutual agreement.

Medical Emergency (EMS) A situation with a real or perceived need for immediate medical care based on injury or other unforeseen acute physical or mental disorder.

Medical Mutual Aid (MMA) (EMS) The provision of medical assistance to a requesting county by another government jurisdiction when the affected county does not have sufficient medical resources to respond effectively to a disaster.

Medical Officer Individual in the command post responsible for the observation and medical treatment of those working at the site of a hazardous materials incident.

Medium Speed Not exceeding thirty miles per hour.

Mega- Prefix meaning one million (1,000,000).

Melting Point Temperature at which a solid is changed to a liquid.

Membrane The impermeable layer of a sanitary landfill liner.

Memorandum of Agreement A written accord between administrative agencies that clarifies or establishes joint procedures or authorities necessary to administer a program.

Memo Waybill Waybill used when agent does not have sufficient information to determine freight charges; contains adequate information to properly handle car.

Memorandum Bill of Lading Duplicate copy of a bill of lading.

Merchandise Car A RR car containing several less-than-carload shipments.

MESA Mining Enforcement and Safety Administration; all respirators designated for use with pesticides are jointly approved by MESA and the National Institute for Occupational Safety and Health (NIOSH).

Metabolite Compound resulting from chemical action on a pesticide inside a living organism; may be more or less toxic than its "parent" pesticide; may develop in some cases when a pesticide is exposed outside a living organism.

Methane Colorless, odorless, combustible gas produced by the decomposition of vegetable matter or by chemical synthesis.

Metric System of measurement used by most of the world; meters (length), grams (weight), and liters (volume).

122

Metric Ton Unit of mass and weight equal to 1,000 kilograms or 2,205 pounds avoirdupois.

mg *see* MILLIGRAM.

mg/kg Milligrams per kilogram; used to express the amount of pesticide in mg/kg of animal body weight that produces a known effect.

mg/l Milligram/liter.

MHz Megahertz; an identification of radio frequencies.

Micro- Prefix meaning 1/1,000,000.

Microbial Pesticide Chemical whose active ingredient is a bacteria, virus, or other tiny plant or animal.

Microgram Metric weight measurement equal to 1/1,000,000th of a gram; approximately 28,500,000 micrograms equal one ounce.

Microorganisms Plant or animal life not visible to the human eye without the aid of a microscope; found in air, water, and soil and generally include the bacteria, yeasts, and fungi.

Microwave Plasma The gas generated during the detoxification reaction in an experimental hazardous waste chemical treatment process by which new stable compounds are synthesized or molecules are decomposed by microwave reactions with gas molecules.

Microwaves Radio frequency waves generated by electronic devices in which electrons are accelerated and directed toward a target.

Midnight Dumper Person or company disposing of hazardous waste illegally (also called Gypsy Dumper).

Mileage Allowance Allowance based on distance made by railroads to owners of private freight cars.

Mileage Rate Rates applicable to distance.

Milli- Prefix meaning 1,000.

Milligram (mg) Weight measurement equal to 1/1,000 gram; approximately 28,500 mg equal one ounce.

Milling in Transit The stopping of grain, lumber, etc., at a point located between the points of origin and destination for the purpose of milling.

Milliroentgen Measure of radiation; 1/1000 of a roentgen.

Mine An excavation made in the earth to extract ores.

Mineral-Based Sorbent   An inorganic substance with adsorptive or absorptive capacities, used to recover oil; includes vermiculite, perlite, or volcanic ash; recovers four to eight times its weight in oil.

Mineral Spirits   Flammable petroleum distillates that boil at temperatures lower than kerosene; used as solvents and thinners especially in paints and varnishes; naphthas; used extensively in chemical dispersants before 1970, not in use now due to their toxicity.

Minimum Charge   Least charge for which a shipment will be handled.

Minimum Rate   Lowest rate that may be charged.

Minumum Weight   Least weight at which a shipment is handled at a carload rate.

Mining Overburden   Material overlying an economic mineral deposit which is removed to gain access to that deposit.

Ministerial   A governmental decision requiring no personal judgment by a public official as to manner of execution of the project; involves only the use of fixed standards or objective measurements without personal or subjective judgment (such as automobile registrations, dog licenses, and marriage licenses).

Miscible Flood   An oil recovery process involving injection of a solvent followed by a displacing fluid.

Miscible Liquids   Two or more liquids that can be mixed and will remain mixed under normal conditions.

Mist   Suspended liquid droplets generated by condensation from the gaseous to the liquid state or by breaking up a liquid into a dispersed state, such as by splashing, foaming, or atomizing. Mist is formed when a finely divided liquid is suspended in air.

Mites   Tiny animals related to insects; have eight jointed legs, two body regions, no antennae, and no wings; have six legs during nymphal stage and are often grouped with ticks and spiders.

Miticide   Pesticide used to control mites and ticks.

Mitigation   To make less harsh; alleviate. In negotiations includes:
1. Avoiding the impact altogether by not taking a certain action
2. Minimizing impacts by limiting the degree or magnitude of the action and its implementation
3. Rectifying the impact by repairing, rehabilitating, or restoring the impacted environment
4. Reducing or eliminating the impact over time by preservation and maintenance operations during the life of the action

5. Compensating for the impact by replacing or providing substitute resources or environments.

Mixed Carload (RR)   Carload of different articles in a single consignment.

Mixed Carload Rate   Freight charge applicable to a carload of different articles in a single consignment.

Mixed Paper   Waste paper of various kinds and quality; usually collected from stores, offices, and schools.

Mixtures   Substances containing two or more materials not chemically united.

mm   Millimeters

Mobile Intensive Care Nurse (MICN)   Authorized Registered Nurse.

Mobile Intensive Care Unit   An emergency vehicle staffed by EMT, mobile intensive care nurses, and/or physicians and equipped to provide intensive care to the ill or injured at the scene of medical emergencies and during transport to a hospital.

Mode of Action   Manner in which an herbicide controls weeds or plant growth; also the ways in which pesticides affect people and other mammals.

Moderator   Material such as ordinary water, heavy water, or graphite, used in a reactor to slow down high-velocity neutrons, thus increasing the likelihood of further fission.

Moisture Content   Amount of water absorbed in a fuel at a given time; has an effect of regulating ignition time and rate of combustion of a fuel. A fuel generally absorbs moisture from the air but can release moisture to drier air, an important factor in fire behavior.

Mold   A fungus-caused growth often found in damp or decaying areas or on living things.

Molded Pulp Products   Contoured fiber products molded from paper pulp for protective packaging, such as egg cartons.

Molecular Weight   Total mass of any group of atoms bound together to act as a single unit.

Molecule   Smallest part of a substance that retains the chemical and physical properties of that substance; a combination of two or more atoms of the same or different elements.

Molluscicide   Pesticide used to control slugs and snails.

Mollusks   Any of a large family of invertebrate animals, including snails and slugs.

125

Monitoring (1) Periodic or continuous determination of radiation levels. (2) Ongoing surveillance of a hazardous materials waste disposal site by measurements or observations of its ambient air, ground water, surface water, and soil conditions.

Monitoring, Environmental Contamination Use of instruments and other techniques to determine the presence or levels of hazardous materials.

Monitoring Equipment Instruments and devices used to identify, quality, and/or quantify contaminants.

Monitoring, Radiological Operation of locating and measuring radioactive contamination by means of survey instruments that can detect and measure (as dose rates) ionizing radiations.

Monitoring System Regular measurements of any pesticide escaping into the environment.

Monitoring Well A well installed to routinely observe ground water levels or to systematically collect water samples and analyze these for chemical pollution.

Mono- A numbering prefix indicating the number one.

Monofills Landfills, surface impoundments, or waste piles used to treat, store, or dispose of one or more of a small group of inorganic wastes; includes wastes that are hazardous solely because they exhibit EP toxicity.

Motor Vehicle Fuels Hazardous materials used primarily to fuel motor vehicles or engines.

Movable Fuel Storage Enders (Farm Carts) Containers in excess of 1200 gallons water capacity equipped with wheels to be towed from one location to another; basically nonhighway vehicles used as fuel supply for farm tractors, construction machinery, and similar equipment.

MPH Miles Per Hour.

MSDS see MATERIAL SAFETY DATA SHEET.

Mucous Membrane The lining of the nose, mouth, eyes, vagina, and rectum. Mucous membranes are not as durable as other skin; contact of infected body fluids or chemicals with intact mucous membranes may transmit disease or cause injury.

Mudflap Deflecting shield at rear of wheels to minimize road surface substances from being thrown rearward; mudflaps designed to remain rigid when vehicle is travelling at high speed are termed "antisail."

126

Mudhop (RR slang)   Yard clerk.

Multi-Hazard Functional Planning   The California format used for developing disaster and emergency plans.

Multiple-Casualty Incident (EMS)   An emergency involving two or more casualties requiring response of more than one medical transport unit with transport priority given to the most seriously injured.

Multipurpose   Capable of more than one use; multipurpose pesticides may control more than one pest.

Multipurpose Staging Area (MSA)   A predesignated location such as a fairground or school facility having large parking areas and shelter for equipment and operators, which provides a base for coordinated localized emergency operations, a rally point for mutual aid coming into an area, and a site for post-disaster population support and recovery activities.

Municipal Solid Waste   Non-hazardous, non-agricultural solid waste generated by residences, businesses, and institutions.

Municipality   A city, borough, incorporated town, township, or county, or any authority created by one of the foregoing.

Mutagenic   Capable of producing a genetic change.

Mutagenesis   Alteration of the inherited genetic material, i.e., alteration of DNA in the paternal or maternal reproductive cell; may cause infant to be born malformed.

Mutual Aid Agreement   An agreement in which two or more parties agree to furnish resources and facilities and to render services to each and every other party of the agreement to prevent and respond to any type of disaster or emergency.

Mutual Aid Region   A subdivision of the state's emergency organization established to facilitate the coordination of mutual aid and other emergency operations within a multicounty geographical area of the state.

MWE   Megawatts of Electricity; rating of power output of a generating station.

Mycoplasma-Like Organisms   Organisms recently discovered to be the cause of many plant diseases formerly attributed to viruses; organisms smaller than bacteria and larger than viruses.

Mylar   A thin plastic material used in alpha survey equipment.

N

Na₂SO₃ Sodium sulfite.

Nano- Prefix that divides a basic unit by one billion; 1/1,000,000,000.

Naptha Various volatile and often flammable liquid hydrocarbon mixtures used as solvents and diluents; consist mainly of hydrocarbons with higher boiling point than gasolines and lower boiling point than kerosene; principal component of chemical dispersants used prior to 1970.

Napthenes Class of hydrocarbons with similar physical and chemical properties to alkanes; insoluble in water, generally boil at 10° to 20°C higher than corresponding carbon number alkanes.

Narrow Gauge (RR track) When the distance between the heads of the rails is less than four feet eight inches.

Narrowleaf Plants, Species, or Weeds Plants having narrow leaves and parallel veins, as compared to broadleaf plants; examples: grasses, sedges, rushes, and onions.

National Contingency Plan (NCP) Created by CERCLA to define the federal response authority and responsibility for oil and hazardous material spills.

National Emergency Equipment Locator System (NEELS) A computerized inventory containing description, location, and who to contact information for emergency spill cleanup equipment in parts of the United States and Canada.

National Fire Protection Association (NFPA) An international voluntary membership organization that promotes improved fire protection and prevention, establishes safeguards against loss of life and property by fire, and writes and publishes the American National Standards.

National Incident Management System (NIMS) A standardized approach to incident management that consists of five major subdivisions collectively providing a total systems approach to all-risk incident management.

National Institute for Occupational Safety and Health (NIOSH) A federal agency that, among other activities, tests and certifies respiratory protective devices, air sampling detector tubes, and recommends occupational exposure limits for various substances.

National Oceanic and Atmospheric Administration (NOAA) A government agency tasked with surveying.

National Pesticide Telecommunications Network (NPTN) A 24-

hour national hotline (1-800-858-PEST) operated by the Texas Tech. University School of Medicine, providing toll-free information about pesticide safety, application, chemistry, and toxicology to callers in the U.S., Puerto Rico, and the Virgin Islands. Questions are answered directly or via the next day's mail.

National Pollutant Discharge Elimination System (NPDES) A program established under the Federal Water Pollution Control Act requiring all point source discharges into any body of water to be permitted by EPA or the designated state agency; minimum pretreatment requirements for such discharges are established under this program.

National Response Center (NRC) A communications center for activities related to response actions, located at Coast Guard headquarters in Washington, DC. The NRC receives and relays notices of discharges or releases to the appropriate (OSC), disseminates OSC and RRT reports to the National Response Team (NRT) when appropriate, and provides facilities for the NRT to use in coordinating a national response action when required. The toll free number (800-424-8802 or 202-426-2675 or 202-267-2675 in the Washington, DC area) can be reached twenty-four hours a day for reporting actual or potential pollution incidents.

Natural Circulation Reactor Nuclear reactor in which the coolant (usually water) is made to circulate without pumping, but by natural convection resulting from the different densities of its cold and reactor-heated portions.

Natural Enemies Predators and parasites in the environment that attack pest species.

Natural Formation An undisturbed material.

Natural Gas A mixture of hydrocarbons and varying quantities of nonhydrocarbons existing in natural underground reservoirs, either in a gaseous phase or in solution with crude oil.

Natural Gas Liquids Those portions of reservoir gas liquefied at the surface in leases separators, field facilities, or gas processing plants.

Natural Hazards Geological, meteorological, or biological conditions that affect the safety of facility operations, thereby posing potential risks to human health and the environment.

Natural Organic Sorbent Natural materials such as peat moss, straw, and sawdust which can be used to recover spilled oil; generally absorb three to six times their weight in oil because of the criss-cross arrangement of fibers within the material. All natural sorbents will absorb water as well as oil and virtually all sink when saturated with water.

Natural Resources Land, fish, wildlife, biota, air, groundwater, drinking water supplies, and other such resources belonging to, managed by, held in trust by, appertaining to, or otherwise controlled by the United States, state or local government, foreign government, or private concern, or individual.

Natural Uranium (Normal Uranium) Uranium as found in nature, containing 0.7% 235U, 99.3% 238U, and a trace of 234U.

NCP National Oil and Hazardous Substances Pollution Contingency Plan (40 CFR Part 300); prepared by EPA to put into effect the response powers and responsibilities created by CERCLA and the authorities established by Section 311 of the Clean Water Act.

NCRPM National Council on Radiation Protection and Measurements; an independent organization chartered by the U.S. Congress.

ND Nondetect.

Necrosis Death in a particular part of a living tissue; example: death of a certain area of a leaf.

Necrotic Showing varying degrees of dead spots or areas.

Negative Declaration A written statement by a lead agency briefly describing reasons that a proposed project, not exempt from an environmental quality agency, will not have a significant effect on the environment and therefore does not require preparation of an environmental impact report.

Negligence Failure to do a job or duty; an act or state of neglectfulness.

Negligible Residue A tolerance set on a food or feed crop containing a very small amount of pesticide at harvest as a result of indirect contact with the chemical.

Negotiation A process wherein parties compromise to reach a mutual agreement.

Nematicide Pesticide used to control nematodes; often applied as a soil fumigant.

Nematode Worm-like organism feeding on or in plants and animals; some are microscopic, some larger; many are internal parasites of people and other animals; common names: roundworms, threadworms, and eelworms.

Neoprene Synthetic rubber used to make gloves and boots offering protection against most pesticides.

Nephratoxic A substance that negatively affects the kidneys.

Nervous System   All nerve cells and tissues in animals, including the brain, spinal cord, ganglia, nerves, and nerve centers.

Nested   Packed one within another.

Net Ton   2,000 pounds.

Net Ton-Mile (RR)   Movement of a ton of freight one mile.

Net Weight   (1) Weight of an article clear of packing and container. (2) As applied to a carload, the weight of the entire contents of the car.

Net, Network (EMS)   An orderly arrangement of stations interconnected through communications channels to form a coordinated entity.

Neurotoxic   Poisonous to nerve cells.

Neutralize   To destroy the effectiveness of a thing; to make harmless anything contaminated with a chemical agent.

Neutralization   Process in which acid or alkaline properties of a solution are altered by the addition of certain reagents to bring the hydrogen and hydroxide concentrations to an equal value; sometimes referred to as 7 pH, the value of pure water.

Neutralization Surface Impoundments   Surface impoundments that: (1) contain no other wastes; (2) are used to neutralize wastes that are hazardous solely because they exhibit the characteristic of corrosivity; (3) neutralize the corrosive wastes sufficiently rapidly so there is no potential for migration of hazardous waste from the impoundment.

Neutron   A component of a nucleus that is 2,000 times more massive than an electron; differs from a proton by its neutral (zero) electrical charge; also a type of ionizing radiation.

Newsprint   Type of paper generally used for printing newspapers.

NFA   The National Fire Academy; a component of FEMA's National Emergency Training Center located in Emmitsburg, Maryland. It provides fire prevention and control training for the fire service and allied services. Courses on campus are offered in technical, management, and prevention subject areas. A growing off-campus course delivery system is operated in conjunction with state fire training program offices.

NFPA   National Fire Protection Association.

NFPA 704M   A pamphlet describing a system for identifying fire hazards, published by the NFPA.

NHMIE   National Hazardous Materials Information Exchange; provides information on hazmat training courses, planning techniques, events and conferences, and emergency response experiences and lessons learned. Call toll free 1-800-752-6367 (in Illinois, 1-800-367-9592). Planners with personal computer capabilities can access NHMIE by dialing FTS 972-3275 or (312) 972-3275.

NHTSA National Highway Traffic Safety Administration of Department of Transportation; responsible for establishing motor vehicle safety standards and regulations for new vehicles; formerly the National Highway Safety Bureau (NHSB).

Nine One One (911) A three-digit emergency telephone number promulgated as the nationwide emergency access number.

NIOSH National Institute for Occupational Safety and Health; all respirators designated for use with pesticides are jointly approved by NIOSH and MESA; see MESA.

NIST National Institute of Standards and Technology.

Nitrophenols Synthetic organic pesticides containing carbon, hydrogen, nitrogen, and oxygen; used as wood preservatives, fungicides, or disinfectants; affect liver and central nervous system in the human body.

NLPGA National LP-Gas Association.

NNI Nederlands Normalisatie-instituut (Netherlands).

NOAA National Oceanic and Atmospheric Administration.

NOAA Weather Station A mobile weather data collection and forecasting facility (including personnel) provided by the NOAA which can be utilized at a hazardous materials incident.

NOAEL No Observable Adverse Effects Level; an estimate of dose or exposure threshold levels for toxic effects. It is determined from observations reported in a specific scientific study; see LOAEL.

NOI Not Otherwise Indexed; legal classification of radioactive material transported in sole-use vehicles.

Nona- (or Ennea-) A prefix indicating the number nine.

Non-Accumulative Does not build up or store in an organism or in the environment.

Non-Agency Station A railroad station that does not have an agent; a closed station.

Non-Destructive Testing Testing to detect internal and concealed defects in materials using techniques that do not damage or destroy the items being tested; X-rays, isotropic radiation, and ultrasonics are frequently used.

Non-Ferrous Metals Metals that contain no iron, such as aluminum, copper, brass, bronze.

Nonflammable Gas A compressed gas not classified as flammable.

Nonliquefied Gas A gas that is entirely gaseous at a temperature of 70°F (21°C).

Non-Persistent Pesticide Pesticide that breaks down almost immediately or only lasts for a few weeks or less and turns into non-toxic byproducts; may be broken down by light, moisture, or microorganisms, or may evaporate.

Non-Point Surface A source from which pollutants emanate in an unconfined and unchannelled manner, including but not limited to (1) water effluent not controlled through NPDES permits or traceable to a discrete identifiable origin but resulting from natural processes, such as nonchannelled runoff, precipitation, drainage, or seepage; (2) air contaminant emissions from landfills and surface impoundments.

Non-Selective Pesticide A pesticide that is toxic to a wide range of pests or toxic to more than one animal or plant; broad spectrum.

Non-Target Any plant, animal, or other organism that is not the object of a pesticide application; any species at which the pesticide is not aimed.

Non-Toxic Not poisonous.

Non-Volatile A substance that does not evaporate at normal temperatures when exposed to the air.

Nor- A prefix indicating the parent compound from which another may be regarded as derived, as norcamphor. Also used to indicate a normal compound which is isomeric, with the one to the name of which it is prefixed.

Normal Speed Maximum authorized speed shown in the timetable.

North American Number (NA) A four-digit number in the United States and Canada to identify a hazardous material (HM) or a group in transportation.

NOS Not Otherwise Specified, a legal classification of radioactive material transported in sole-use vehicles.

Nosocomial "Originating in the hospital;" a disease spread by contact with the health-care system.

Noxious Weed A plant considered to be especially unwanted or troublesome.

Nozzles Devices that control drop size, rate, uniformity of coverage; in use with a pesticide, they also determine the thoroughness, and safety of a pesticide application. Nozzle type determines the ground pattern of coverage.

133

NPDES   National Pollution Discharge Elimination System.

NRT   National Response Team; consisting of representatives of fourteen government agencies (DOD, DOI, DOT/RSPA, DOT/USCG, EPA, DOC, FEMA, DOS, USDA, DOJ, HHS, DOL, Nuclear Regulatory Commission, and DOE), is the principal organization for implementing the NCP. When the NRT is not activated for a response action, it serves as a standing committee to develop and maintain preparedness, to evaluate methods of responding to discharges or releases, to recommend needed changes in the response organization, and to recommend revisions to the NCP. The NRT may consider and make recommendations to appropriate agencies on the training, equipping, and protection of response teams; and necessary research, development, demonstration, and evaluation to improve response capabilities.

NSAI   National Standards Authority of Ireland (Ireland).

NSF   Norges Standard Iseringsforbund (Norway).

NSF   National Strike Force; made up of three strike teams. The United States Coast Guard, counterpart to the EPA's Environmental Response Teams.

NTU   Nephelometric Turbidity Unit.

Nuclear   An adjective referring to the atom's nucleus.

Nuclear Battery   A radioisotopic generator.

Nuclear Chain Reaction   A sequence of fissions in which neutrons, released during the spontaneous disintegration of a nucleus, collide with other nuclei and cause these in turn to disintegrate and release more neutrons.

Nuclear Energy   Energy liberated by a nuclear reaction (fission or fusion) or by radioactive decay.

Nuclear Explosive   An explosive based on fission or fusion of atomic nuclei.

Nuclear Power Plant   Any device, machine, or assembly that converts nuclear energy into some form of useful power, such as mechanical or electrical; in a nuclear electric power plant, heat produced by a reactor is generally used to make steam to drive a turbine that in turn, drives an electric generator.

Nuclear Reaction   Change in an atomic nucleus, such as fission, fusion, neutron capture, or radioactive decay, as distinct from a chemical reaction, which is limited to changes in the electron structure surrounding the nucleus.

Nuclear Reactor   A device in which a fission chain reaction can be initiated, maintained, and controlled; essential component is a core with fissionable fuel; usually has a moderator, a reflector, shielding, coolant, and control mechanisms.

Nuclear Transformation   The change in an isotope from a less stable to a more stable form, typically involving the emission of alpha, beta, gamma, and/or neutron radiation.

Nuclear Weapon   A device capable of producing a nuclear yield by either fission, fusion, or both; contains conventional explosives.

Nucleonics   The science and technology of nuclear energy and its applications.

Nucleus   The center part of an atom, composed of protons and neutrons; plural: nuclei.

Nuclide   A general term applicable to all atomic forms of the elements; term is often erroneously used as a synonym for "isotope," which has a more limited definition. Whereas isotopes are the various forms of a single element (so are a family of nuclides) and all have the same atomic number and number of protons, nuclides comprise all the isotopic forms of all the elements. Nuclides are distinguished by their atomic number, atomic mass, and energy state.

NUREG 0654/FEMA-REP-1   Criteria for Preparation and Evaluation of Radiological Emergency Response Plans and Preparedness in Support of Nuclear Power Plants, prepared by NRC and FEMA; provides a basis for state and local government and nuclear facility operators to develop radiological emergency plans and improve emergency preparedness. The criteria also will be used by federal agency reviewers in determining the adequacy of state, local, and nuclear facility emergency plans and preparedness.

Nymph   The stage of development in certain insects after hatching when the young insect looks like an adult but lacks fully developed wings.

O

O_2 Oxygen

Objective   The main purpose to be achieved by tactical units at a hazardous materials emergency.

Obsolete Scrap   Scrap materials derived from products that have completed their useful economic life.

Occupational Exposure   Limited to infectious disease exposure; reasonably anticipated skin, eye, mucous membrane, or parenteral contact with blood or other potentially infectious materials that may result from the performance of an employee's duties (OSHA). This definition excludes incidental exposures that may take place on the job, that are neither reasonably or routinely expected, and that the worker is not required to incur in the normal course of employment; see EXPOSURE DETERMINATION or EXPOSURE INCIDENT.

Occupational Safety and Health Administration (OSHA) Component of the United States Department of Labor, an agency with safety and health regulatory and enforcement authorities for most United States industry, business, and states.

Ocita A prefix indicating the number eight.

Odor Threshold The lowest concentration in the atmosphere that can be detected by the nose.

Office Car Car used by railway officials while traveling.

Office of Emergency Services (OES) A California state agency responsible for assisting local jurisdictions in preparing for, responding to, and recovering from the effects of a disaster.

Office of Hazardous Materials Regulation (OHMR) A federal agency tasked with enforcement of transportation regulations set forth in CFR 49.

Office of Hazardous Materials Transportation (OHMT) A federal agency tasked with the research and recommended revisions to CFR 49.

Off-Scene Support Assistance (via telephone, radio, or computer) from technical persons, agencies, shippers, responders, etc., not at the accident site.

Off-Site Facility A recycling, treatment, or disposal facility located at a site other than where the wastes are generated; often operated by a firm whose business is solely the processing of industrial wastes and which itself generates little or no hazardous waste.

Off-Site Hazardous Waste Facility An operation involving handling, treatment, storage, or disposal of hazardous wastes such that the waste is transported commercially to the site.

OHMTADS Oil and Hazardous Materials Technical Assistance Data System; a computerized data base containing chemical, biological, and toxicological information about hazardous substances. On-scene coordinators use OHMTADS to identify unknown chemicals and to learn how to best handle known chemicals.

-oic A suffix indicating a carboxylic acid, such as pentanoic acid. Often the suffix -ic is used in forming the name of an acid like acetic acid.

136

Oil Oil of any kind, including crude, refined, petroleum, fuel oil, animal oil, sludge, oil refuse, and oil mixed with wastes.

Oil Hazardous Materials Technical Assistance Data System Organization within the EPA that provides information to emergency teams responding to spills on some hazardous substances.

Oil-in-Water Emulsion An emulsion of oil droplets dispersed in surrounding water, formed as a result of wave action or by a chemical dispersant; the oil droplets show a tendency to coalesce and reform into an oil slick when the water becomes calm.

Oil Spill Cleanup Agent Agents used in removing oil from the environment, including inert sorbent materials, approved chemical dispersants, surface collecting agents, sinking agents, and biological additives.

Oils Liquids used to carry an active ingredient in a pesticide formulation, to weaken a pesticide formulation, or for direct application as a pesticide.

Oil Slug A downward-moving oil mass, which often results when oil is spilled on relatively porous soil; the slug-like shape results from the tendency of the mass to leave behind a funnel of soil which is partially saturated with oil.

Oil Spill Contractor Private firms that have been formed to provide oil spill cleanup services; often have their own containment and recovery equipment and may be identified in local and regional contingency plans.

Oil Spill Cooperative Organizations formed by oil companies operating in a given area for the purpose of pooling equipment and training personnel to combat oil spills.

-ole A suffix used for phenolic ethers, such as phenetole, and for aldehydes, such as oenathole.

Oleophilic Agent A material or chemical that has the tendency to attract oil; chemicals of this type may be used to treat sorbent materials in order to increase their oil recovery capacity.

Olfactory Pertaining to the sense of smell.

ON Osterreichisches Normungsinstitut (Austria).

-one A suffix usually indicating the presence of a ketone group as in propanone or acetophenone.

On the Ground (RR) On the ties, not on the rails of a railroad track; example: a derailed train.

On-Scene Commander (OSC) The overall coordinator of an oil spill response team, usually a representative of an oil company, a government official, or an independent oil spill cleanup contractor;

responsible for onsite strategical decisions and actions throughout each phase of a cleanup operation; maintains close liaison with the appropriate government agencies to obtain support and provide progress reports on each phase of the emergency response.

On-Site Facility A treatment process or facility located on or adjacent to the premises of a waste generating firm and usually operated by that firm.

Open Burning The burning of waste materials in the open or in a dump; produces smoke, odor, and other objectionable air pollutants.

Open Cycle Reactor System A reactor system in which the coolant passes through the reactor core only once and is then discarded.

Open Dump An open land site where solid waste is handled and disposed of improperly; usually the site is exposed to the elements, harbors insects and rodents, and leaks methane gas. The Federal Resource Conservation and Recovery Act (RCRA) authorizes the State Solid Waste Management Board to inventory open dumps in California as a step in enforcing minimum waste disposal standards.

Open Dump Inventory A survey undertaken by the State Solid Waste Management Board's enforcement division, in cooperation with the federal Environmental Protection Agency, to determine the number, location, and regulation compliancy of all disposal facilities and open dumps in California; this inventory will help the Board put an end to improper solid waste disposal practices.

Operator The person responsible for the overall operation of a facility.

Operating Speed The constant rate at which a pesticide sprayer moves during application; usually measured in miles per hour or feet per minute.

Operational Period (ICS) A period of time scheduled for execution of a given set of suppression and rescue actions as specified in the Incident action plan.

Operational Radiation Monitoring Systems (ORMS) An instrumentation system at a nuclear reactor designed to detect and alarm abnormal radiation levels in process and effluent streams.

Operations Organizational level within the Incident Command System (ICS) immediately subordinate to the incident commander. This position is responsible for the direct management of all incident tactical activities. National Incident Management System (NIIMS).

Opposing Signals Signals that govern movements in opposite directions on the same railroad track.

138

Oral   Of, by, through, or into the mouth.

Oral Toxicity   How poisonous a pesticide is to an animal or person when taken by mouth.

Orange Oxide   Uranium trioxide.

Organic   Chemical compound containing carbon; may occur in nature while others are produced by chemical synthesis.

Organic Chemistry   The study of the compounds of carbon.

Organic Cooled Reactor   A reactor that uses organic chemicals such as mixtures of polyphenyls (diphenyls and terphenyls) as coolant.

Organic Matter   Chemical substances containing the element carbon, originating in animal or plant life or in their derivatives, coal or petroleum.

Organic Peroxide   An organic derivative of the inorganic compound hydrogen peroxide.

Organic Pesticides   Pesticides that contain carbon; major groups are petroleum oils and synthetic organic pesticides.

Organic Phosphate Insecticide   Pesticide, examples of which include parathion, HETP, and TEPP.

Organism   Any living thing.

Organochlorine Compounds   Synthetic organic pesticides that contain chlorine, carbon, and hydrogen; affect the central nervous system.

Organophosphates   Synthetic organic pesticides that contain carbon, hydrogen, and phosphorous; highly toxic to humans as they prevent proper transmission of nerve impulses.

Organs   Body parts affected by substance exposure.

Original Container   The package (bag, can, bottle, etc.) prepared by the manufacturer in which the pesticide is placed and then sold; package must be labeled with what the pesticide is, how to use it safely and correctly, and how to legally dispose of the empty container.

ORMs   see OTHER REGULATED MATERIALS.

OSC   On-Scene Coordinator; the federal official predesignated by EPA or USCG to coordinate and direct federal response and removals under the NCP; the DOD official designated to coordinate and direct the removal actions from releases of hazardous substances, pollutants, or contaminants from DOD vessels and facilities. When the NRC receives notification of a pollution incident, the NRC duty officer notifies the appropriate OSC, depending on the location of an incident. Based on this initial report and any other information that can be obtained, the OSC makes a preliminary assessment of the need for a federal response. If an on-scene response is required, the OSC will go to the scene and monitor the response of the responsible party or state or local government. If the responsible party is unknown or not taking appropriate action, and the response is beyond the capability of state and local governments, the OSC may initiate federal actions, using funding from the FWPCA Pollution Fund for oil discharges and the CERCLA Trust Fund (Superfund) for hazardous substance releases.

-ose   A suffix usually indicating a carbohydrate such as glucose or cellulose. Also used in naming the primary alteration of hydrolytic products of proteins, such as albumose.

OSHA   Occupational Safety and Health Administration.

Osmosis   The tendency of a fluid to pass through a semipermeable membrane typically separating a solvent and a solution so as to tend to equalize their concentrations on both sides of the membrane.

Other Regulated Materials (ORM)   Materials that do not meet the definitions of hazardous materials, but possess enough hazardous characteristics that they require some regulation; relating to ORM-A, -B, and -C; ORM-D materials are hazardous materials transported in small quantities.

ORM A—A material that has an anesthetic, irritating, noxious, toxic, or other similar property and which can cause extreme annoyance or discomfort to passengers and crew in the event of leakage during transportation [CFR 49, 173.500(b)(1)].

ORM B—A material (including a solid when wet with water) capable of causing significant damage to a transport vehicle from leakage during transportation [CFR 49, 173.500(b)(2)].

ORM C—A material that has other inherent characteristics not described as an ORM A or ORM B but which make it unsuitable for shipment, unless properly identified and prepared for transportation [CFR 49, 173.500(b)(4)].

ORM D—A material such as a consumer commodity which, though otherwise subject to the regulations of this subchapter, presents a limited hazard during transportation due to its form, quantity, and packaging [CFR 49, 173.500(b)(4)].

ORM E—A material that is not included in any other hazard class, but is subject to the requirements of CFR 173.500, and includes hazardous waste.

Outcome   Result.

Outlet Valve   The valve farthest downstream in a tank piping system to which the discharge hose is attached.

Out-of-Service Resources (ICS) Material or personnel resources assigned to an incident but unable to respond for mechanical, rest, or personnel reasons.

Outrigger Structural load-carrying members attached to and extending outward from the main longitudinal frame members of a trailer.

OVA Organic Vapor Analyzer.

Overhead Personnel Supervisory personnel.

Over the Top Pesticide application over the top of a growing plant.

Overpack An enclosure used to consolidate two or more packages of hazardous material; does not include a freight container.

Overpressure (1) The transient pressure over and above atmospheric pressure caused by a shock wave from a nuclear explosion. (2) Pressure in excess of the design or operating pressure.

Overstress To stress an object or person beyond recoverable limits.

Overturn Protection for fittings on top of a tank in case of rollover; may be combined with flashing rail or flashing box.

Ovicide Pesticide used to control eggs, particularly insect, mite, or nematode eggs.

Owner The person or municipality who is the owner of record of a facility or part of a facility.

Oxidation (1) The combining of oxygen with another substance chemically; when process is slow, no apparent heat or light is produced, as in rusting of metal or yellowing of paper; when rapid, oxidation is accompanied by the production of heat and light (as in fire). (2) In hazardous waste management, a process to treat a waste stream with a strong oxidizing agent changing the waste chemically to a less hazardous state.

Oxidizer Substance that yields oxygen readily to stimulate the combustion of organic matter and inorganic matter.

Oxidizing Ability The ability to yield oxygen readily to stimulate combustion.

Oxidizing Agents Substance containing chemically available oxygen.

Oxygen An element that readily unites with materials.

Oxygen Deficiency A concentration of oxygen insufficient to support life.

Oxygen-Deficient Atmosphere An atmosphere that contains an oxygen content of 19.5 percent or less by volume at sea level.

Pacific Strike Team The National Strike Force pollution control team equipped and trained to assist in the response to oil or chemical incidents occurring in the western United States and administered by the United States Coast Guard.

Package Car RR car containing several less-than-carload shipments.

Package Freight Merchandise shipped in less-than-carload quantities.

Package Markings The descriptive name, instructions, cautions, weight or specification marks required to be placed upon the outside of containers of hazardous materials.

Package Power Reactor A small nuclear power plant designed to be crated in packages small enough to be conveniently transported to remote locations.

Packaging A broad term used by the Department of Transportation to describe shipping containers and any markings, labels, or placards affixed to them.

Packaging Materials Any of a variety of papers, cardboards, metals, wood, paperboard, and plastics used in the manufacture of containers for food, household, or industrial products.

Pair Production The transformation of the kinetic energy of a high-energy photon or particle into mass, producing a particle and its antiparticle, such as an electron and a positron.

Pallet A small portable platform for holding material for storage or transportation.

Pantograph A device located on top of electric RR equipment that collects power from an overhead contact wire by means of a sliding contact shoe.

Paper A thin sheet material made of cellulose pulp, derived mainly from wood, rags, and certain grasses, processed into flexible leaves or rolls by deposit from an aqueous suspension, used chiefly for writing, printing, wrapping and sanitary purposes.

Paperboard Heavier in weight, thicker, and more rigid than paper. Basic classes of paperboard are container board, box board, and special types such as automobile board, building board, etc.

Paper Rate A published rate under which no RR traffic moves.

Paperstock General term used to designate waste papers that have been sorted or segregated at the source into various grades.

Par- A prefix used to indicate that the substance is related in some way, as a polymer, isomer, etc., of the compound to whose name it is attached, such as paraldehyde, paralactic acid.

Paraffin A waxy substance obtained from the distillation of crude oils; complex mixture of higher carbon number alkanes that is resistant to water and water vapor and is chemically inert.

Parasite An organism that lives and feeds on or in another plant or animal (known as the host) which is usually harmed by the parasite. Parasites are usually pests, but when used to attack other pests which can injure crops or animals, they become forms of biological control.

Parasitic Living off of or feeding on or in another organism.

Parenteral Exposure Communicable disease exposure that occurs through a break in the skin barrier; includes injections, needle sticks, human bites, and cuts contaminated with blood.

Partial Body Exposure The exposure to radiation of only an isolated part of the body such as an arm, foot, or head, rather than the whole body.

Partial Closure Securing a portion of a facility when it is filled to capacity to prevent environmental damage while the remainder of the site is being filled.

Participating Carrier (Tariff) A railroad that is a party, under concurrence, to a tariff issued by another railroad or by a tariff publishing agent.

Particle A minute constituent of matter, generally one with a measurable mass.

Particulate Matter Normally refers to dust and fumes; travels easily through air.

Parts per Billion A unit for measuring the concentration of a particular substance.

Parts per Million Permissible pesticide tolerance on crops is expressed in parts per million (ppm); one ppm is equal to one inch in sixteen miles.

Pathogen Any disease-producing organism. Pathogens can be bacteria, fungi, parasites, or viruses.

Pawl Ratchet; on RR brake wheels, a pivoted bar adapted to fall into the notches or teeth of a wheel as it rotates in one direction and to restrain it from backward motion.

PCB *see* POLYCHLORINATED BIPHENYLS.

PCB-Contaminated Electrical Equipment   Any electrical equipment, including transformers, that contains at least 50 ppm but less than 500 ppm PCB (Title 40 CFR 761.3).

PCB Item   An item containing PCBs at a concentration of five ppm or greater (Title 40 CFR 761.3).

PCB Transformer   Any transformer that contains 500 ppm PCBs or greater (Title 40 CFR 761.3).

Pebble Beach   A beach substrate composed primarily of gravel; more coarse than sand and can allow stranded oil to penetrate to a considerable depth.

Penetrant   An adjuvant that helps a liquid pesticide get into the pores of or through the outer surface of a leaf or root and into the plant; also called a penetrating agent.

Penetration   The ability to get through; the process of entering.

Penta-   A prefix indicating the number five.

Peptization   A method of getting substances into colloidal suspension by breaking down larger particles.

Percent by Weight   A percentage expressing the active ingredient weight as a part of the total weight of the formulation.

Percent Concentration   The weight or volume of an active ingredient expressed as a percentage in the final formulation.

Perched Aquifer   A body of water or water formation located above an impermeable geological formation.

Percolation   The movement, flow, or infiltration of water through the pores or spaces of rock or soil.

Per Diem   A charge made by one transportation line against another for the use of its cars; based on a fixed rate per day.

Perforated Casing   Well casings with holes or slots permitting the passage of fluids or vapors.

Periodic Table (Periodic Chart)   A list of all elements arranged in order of increasing atomic numbers and grouped by similar physical and chemical characteristics into "periods"; based on the chemical law that physical or chemical properties of the elements are periodic functions of their atomic weights.

Perishable   Commodities easily spoiled or damaged by weather or delay in transit; usually describing foodstuffs.

Permeability   The property of soil or rock allowing passage of water through it; depends not only on the volume of openings and pores, but also on how these openings are connected to each other.

144

Permeable   Open to passage or penetration; used especially for a substance that allows the passage of fluids.

Permeation   The molecules of the chemical traveling through the protective material; cannot be seen with the naked eye, but can easily add up to many drops per hour and cause serious skin contact.

Permeation Kits   Kits assembled for the purpose of on-site testing of an unknown liquid substance for permeability of chemical protective clothing.

Permissible Exposure (Maximum)   That dose of ionizing radiation established by competent authorities as an amount below which there is no reasonable expectation of risk to human health, and which at the same time is somewhat below the lowest level at which a definite hazard is believed to exist.

Permit Streamlining Act (A.B. 884)   A California state act, enacted in 1977, that imposes time frames and requirements on governmental-agency permitting processes for development projects.

Permit to Operate   An authorization, issued by Air Pollution Control Districts and Air Quality Management Districts, that is required before operation of a facility and is contingent upon a demonstration that the facility can comply with applicable rules and regulations and with conditions imposed in the Authority to Construct.

Permits   Official approval and permission to proceed with an activity controlled by the permitting authority.

Persist   To remain or stay.

Persistent   When a pesticide remains in the environment for a fairly long time.

Persistent Toxic Substance   A toxic substance that resists natural degradation or detoxification and may present long-term health and environmental hazards.

Personal Alarm Dosimeter   A type of dosimeter that emits an audible alarm whenever a preset threshold value of radiation is reached.

Personal Protective Clothing   Items of attire recommended for protection against hazardous materials, including protective coats, trousers, boots, face and head masks, and gloves.

Personal Protective Equipment (PPE)   The equipment provided to shield or isolate a person from the chemical, physical, and thermal hazards that may be encountered at a hazardous materials incident. Adequate personal protective equipment protects the respiratory system, skin, eyes, face, hands, feet, head, body, and hearing. Per-

145

sonal protective equipment includes both personal protective clothing and self-contained positive-pressure breathing apparatus (NFPA 472, 1–3).

Personnel Decontamination Removal of radioactive contamination from individuals; removal of hazardous substance from individuals.

Personnel Shield A recognizable structure on transportation vehicles that keeps people from accidentally getting too close to the radioactive cargo, thus protecting them against exposure.

Perturbation An event or unexpectedly encountered condition capable of throwing an activity into disorder.

Pest An unwanted organism; a living thing that competes with people for food and fiber or attacks people directly.

Pesticide A chemical or mixture of chemicals used to destroy, prevent, or control any living organism considered to be a pest; any substance used to control or destroy insects, weeds, rodents, disease-causing organisms, and other types of pests that attack living things or spread disease among plants and animals.

Pesticide Chemical A pesticide that is a chemical rather than a nonchemical pesticide (such as a parasite or virus).

Pesticide Kill The death of a large number of nontarget organisms due to careless or improper application or use of a pesticide.

Pesticide Tolerance Amount of pesticide residue that may legally remain on or in a food crop, as established by the EPA; see RESIDUE.

Petroleum Oils Pesticides refined from crude oil that are used to control insects and plants.

Petroleum Products Pesticides and other substances containing oil, gasoline, kerosene, or similar products.

pH A numerical designation of relative acidity and alkalinity; a pH of 7.0 indicates precise neutrality; higher values indicate increasing alkalinity and lower values indicate increasing acidity.

Phase I RCRA Regulations promulgated in May 1980 including the identification and listing of hazardous waste, standards for generators and transporters of hazardous wastes, standards for owners and operators of facilities that treat, store, or dispose of hazardous waste; requirements for obtaining hazardous waste facility permits, and rules governing delegation of authority to the states. see RCRA.

Phase II RCRA Technical requirements for permitting hazardous waste facilities; sets specific standards for particular types of facilities to ensure the safe treatment, storage, and disposal of hazardous waste on a permanent basis by methods that will protect human health and the environment; enable facilities to move from interim status to final facility permits. see RCRA.

Pheromones Chemicals produced by insects and other animals to communicate with other members of the same species; some are used to monitor insect populations; most pheromones used today are synthetic.

Phlebotomist Any health-care worker who draws blood samples.

Photodegradable Able to decompose through a chemical reaction initiated by direct exposure to the sun's ultraviolet light.

Photo-Ionizing Detector A detector used to identify chemical compounds based on their ionization potential.

Photon A minute massless packet of pure light energy; all electromagnetic radiation, including gamma radiation, is in the form of photons.

Physical Description A brief summarization of substance form specifying whether it is solid, powder, flakes, crystals, liquid, gas, etc., accompanied by identification of color and odor where applicable.

Physical Properties Properties of a material relating to the physical states common to all substances, i.e., solid, liquid, or gas.

Physical Treatment The process by which waste is rendered nonhazardous by physically removing the hazardous substance from the waste stream, or is rendered more readily disposable or transportable by reducing the water content or solidifying the waste. Current methods of physical treatment are:

• SEDIMENTATION—removing suspended solid particles by providing time and space for settling in special tanks or holding ponds

• FILTRATION—separating liquids and solids by using various types of filters

• SOLAR EVAPORATION—reducing liquid wastes in volume through simple evaporation in uncovered impoundment ponds

• DISTILLATION—separating liquids with different boiling points by heating the mixture to vaporize and retrieve certain components

Phytotoxic Poisonous to plants.

Pick-Up Plate A sloped structure located forward of the kingpin of a trailer, designed to facilitate engagement of the fifth wheel to the kingpin.

Pico- Prefix meaning one-trillionth; 1/1,000,000,000,000.

147

PID   Photoionization Detector.

Piezometer Nest   Multiple well completions in the same borehole with each well screened over a different interval.

Pig   (1) A Type B container; usually lead, used to ship or store radioactive materials; thick walls provide protection for handlers. (2) A device inserted into pipelines to either clean them out or conduct inspections of pipe condition: pigs that contain instrumentation are called "Smart Pigs."

Piggy-Back   A type of shipping in which bulk containers from one mode, such as highway transportation, are placed on flat cars or container ships for transportation by another mode, such as rail or marine.

Pile   (1) A noncontainerized accumulation of solid, non-flowing hazardous waste. (2) Old term for nuclear reactor, as first reactor was built by piling up graphite blocks and natural uranium.

Pilot (RR)   Employee assigned to a train when the engineman or driver of a track car is not qualified on the physical characteristics or rules of the railroad.

Pin Puller (slang)   A trainman who uncouples cars while switching by lifting the coupler pin with the uncoupling lever located on each end of a car.

Pintle   Hitch.

Pipeline   A product discharge line.

Pipeline Oil   Oil clean enough to be acceptable for transport or purchase.

Piping, Piping Systems   Pipe, tubing, hose, and flexible rubber or metallic hose connectors made up with valves and fittings into complete systems for conveying substances at various pressures in either liquid or vaporized state from one point to another.

Piscicide   Pesticide used to control fish.

Piston Travel (RR Air brake)   The amount of piston movement when forced outward as the brakes are applied.

Pivot Pin   Coupler.

Placards   (1) Diamond-shaped markers 10-3/4 inches square required on a transportation vehicle such as a truck or tank car or a freight container 640 cubic feet or larger. (2) Diamond-shaped sign required on outside of vehicles transporting radioactive materials displaying same standard warning terms and symbols as a label. (3) Paper forms of various designs used to identify RR cars requiring special attention (dangerous, explosives, etc.).

Plane Source   Layer of radioactive sources spread over a given area.

Planning Meeting (ICS)   A consultation held as needed throughout the duration of an incident to select specific strategies and tactics for incident control operations and for service and support planning.

Planning Section Chief (ICS)   Individual responsible for comprehending current situation, predicting probable course of the incident, preparing primary and alternate strategies for the incident commander, and collecting, evaluating, and disseminating information about the incident.

Plant Factor   Ratio of the average power load of an electric power plant to its rated capacity; also called the capacity factor.

Plasma   An electrically neutral gaseous mixture of positive and negative ions; sometimes called the "fourth state of matter," since it behaves differently from solids, liquids, and gases. High-temperature plasmas are used in controlled fusion experiments.

Plastics   Man-made materials consisting of large molecules called polymers containing primarily carbon and hydrogen with lesser amounts of oxygen or nitrogen, frequently compounded with various organic and inorganic compounds as stabilizers, colorants, fillers, and other ingredients.

Plowshare   The Atomic Energy Commission program of research and development on peaceful uses of nuclear explosives; possible uses include large-scale excavation such as for canals and harbors, crushing ore bodies, and producing heavy transuranic isotopes. Term is based on a Biblical reference: Isaiah 2:4.

Plug Back   To shut off lower formation in a well bore.

Plug Door   Door on a refrigerated RR car or box car that is flush with side of car when closed; to open, it is swung out and rolled to one side: also called a sliding flush door.

Plugging and Patching Kits   Kits commercially available or privately assembled for the purpose of providing capabilities for the emergency plugging and patching of leaking containers, pipes, and tanks.

Plume   (1) The column of noncombustible products emitted from a fire (smoke). (2) A vapor cloud formation having shape and buoyancy. (3) The airborne radioactive material released from a nuclear power plant and carried by the prevailing winds that may affect radiologically those downwind areas over which it passes.

Plume Exposure Pathway   Route by which the radioactive material released from a nuclear facility may expose the population-at-risk to radiation; exposure may be external from the passing plume,

149

from contaminated surfaces, or from inhalation of the passing plume.

Plunger Lift   A method of lifting (oil) using a swab or free piston propelled by compressed gas from the lower end of the tubing string to the surface.

Plutonium   A heavy radioactive, synthetic metallic element with atomic number 94; used for reactor fuel and in weapons.

Pneumonitis   Inflammation of the lungs characterized by an outpouring of fluid in the lungs.

Point of Origin   (1) (RR) Station at which a shipment is received by the railroad or freight handler from the shipper. (2) In a fire, the starting point of ignition.

Point of Transfer (LP-Gas)   Location where connections are made or where LP-Gas is vented to the atmosphere in the course of transfer operations.

Point Source   (1) One piece of material that is emitting all the radiation in an area; distinguished from isolated source and plane source. (2) A discernible, confined and discrete conveyance including but not limited to a pipe, ditch, channel, tunnel conduit, well, discrete fissure, container, rolling stock, concentrated animal feeding operation, or vessel or other floating craft from which pollutants are or may be discharged. Does not include return flow from irrigated agriculture.

Poison   A substance that through its chemical action kills, injures, or impairs an organism. Poisons are divided into the following classes:

• CLASS A—Poisonous gases or liquids of such a nature that a very small amount of the gas or vapor of the liquid mixed with air is dangerous to life. (Title 49 CFR 173.326)

• CLASS B—Substances, liquids, or solids other than Class A or irritating materials, which are known to be so toxic to man as to afford a hazard to health (49 CFR 173.343).

Poison Control Centers (PCC)   Certified and designated regional poison control centers which provide information regarding immediate health effects, scene management, victim decontamination, and other emergency medical treatment advice for hazardous materials emergencies.

Poisonous Bait   A food or other substance mixed with a pesticide so that a pest will be attracted to it, will eat it, and will then be killed by it.

Polarized Couplers   Fittings for connecting air brake lines between

150

vehicles; service and emergency couplings are unilateral and will not mate with each other.

Pollinators   Bees, flies, and other insects which visit flowers and carry pollen from flower to flower.

Pollutant   A harmful chemical or waste material discharged into the water, soil, or atmosphere; an agent that makes something dirty or impure.

Pollute   To add an unwanted material (often a pesticide) which may do harm or damage; contaminate, make unclean, or unsafe for use.

Pollution   Contamination of air, water, land, or other natural resources that will or is likely to create a public nuisance or to render such air, water, land, or other natural resources harmful, detrimental, or injurious to public health, safety, or welfare or to domestic, municipal, commercial, industrial, agricultural, recreational, or other legitimate beneficial uses, or to livestock, wild animals, birds, fish, or other life.

Poly-   A prefix indicating polymer of the compound to the name of which it is prefixed, for example polystyrene.

Polychlorinated Biphenyls (PCBs)   A series of hazardous chemical compounds that have been manufactured for more than forty years for such common purposes as electrical insulation and heating/cooling equipment. Now suspected to be carcinogens, PCBs have been disposed of in the air, on land, and in water; recent surveys have detected the presence of PCBs in every part of the country, even those remote from PCB manufacturers.

Polyethylene   A polymer taking the form of a lightweight thermoplastic with high resistance to chemicals, low water absorption, and good insulating properties; manufactured in a number of forms and have been used with considerable success as a sorbent for oil spill cleanup.

Polymerization   Reactions in which two or more smaller molecules chemically combine to form larger molecules; reaction is often violent.

Polyurethane   Any of a class of synthetic resinous, fibrous, or elastomeric compounds belonging to the family of organic polymers, consisting of large molecules formed by the chemical combination of successive smaller molecules into chains or networks; best known are the flexible foams used as upholstery material and mattresses.

Polyvinyl Chloride (PVC)   Common plastic material which, when burned, releases toxic hydrochloric acid.

Pool Car (RR)   Specially equipped cars of different ownerships assigned to a specific company or location.

Pool Reactor   A reactor in which the fuel elements are suspended in a pool of water that serves as the reflector, moderator, and coolant; used for research and training.

Population-at-Risk   Those persons for whom protective actions are being planned, taken, or would be taken.

Porosity   The percentage that the volume of the pore space bears to the total bulk volume; in sand or sandstone, determines the amount of space available for storage of fluids.

Port   An opening for access; in nuclear reactors, an opening through which objects are inserted for irradiation or from which beams of radiation emerge for experimental use.

Portable Container (LP-Gas)   A vessel designed to be readily moved, as distinguished from containers designed for stationary installations.

Portable Storage Container (LP-Gas)   Vessel similar to, but distinct from, those designed and constructed for station installation; designed so it can be readily moved over highways substantially empty of liquid from one usage location to another; has legs or other supports attached or are mounted on running gear (trailer) with suitable supports permitting it to be placed or parked in a stable position on a reasonably firm surface. For large volume, limited duration product usage (such as at construction sites and normally for twelve months or less) these vessels function in lieu of permanently installed stationary containers.

Portable Tank (Skid Tank) (LP-Gas)   Container of more than 1,000 pounds water capacity used to transport LP-Gas handled as a "package," i.e., filled to its maximum permitted filling density; mounted on skids or runners; have all container appurtenances protected in such a manner that they may be safely handled as a "package."

Port of Entry   Location at which foreign goods are admitted into a receiving country.

Positron   A particle also called a beta particle, produced when a proton transforms to a neutron; identical to an electron except for its opposite electrical charge of $+1$.

Post   A vertical structural member.

Post-Consumer Scrap   Any uncontaminated packaging material that is recoverable, such as tin cans, egg cartons, glass bottles/jars.

Post-Emergent   Pesticide used to control a crop or weed after it has appeared.

Post-Emergency Response Portion of an emergency response performed after the immediate threat of a release has been stabilized or eliminated and cleanup of the site has begun.

Post-Incident Analysis The termination phase of an incident that includes completion of the required forms and documentation in conducting a critique.

Post-Remedial Monitoring Report Report covering work done after cleanup completion.

Potency Chemically or medicinally effective; strength; rate of toxicity.

Potential Energy Energy derived from position rather than motion, with reference to a specified field of force.

Potential Test A measurement of the maximum rate at which something can be utilized.

Potentiate To activate.

Potentiation The increase in toxicity (usually considered an undesirable effect) of a pesticide when combined with one or more pesticides.

Potentiometric Surface The surface that represents the level to which water will rise in tightly cased wells.

Pour-On A pesticide that is poured along the midline of the backs of livestock.

Pour Point The lowest temperature at which a substance, such as oil, will flow under specified conditions; important in terms of cleanup since free-flowing oils rapidly penetrate most substrates, whereas semi-solid oils tend to be deposited on the surface and will only penetrate if material is coarse or the ambient temperature high.

Power Density Rate of heat generated per unit volume of a reactor core.

Power Reactor A nuclear reactor designed to produce useful nuclear power, as distinguished from reactors used primarily for research or for producing radiation or fissionable materials.

Pozzolanic Chemical reaction between a silica-alumina-containing solid which produces a concrete-like structure from the separate particles, such as sand and lime.

ppb Parts per billion.

ppm Parts per million.

Pradicide Pesticide used to control vertebrate pests.

Precious Metals   Recoverable gold, silver, or platinum.

Precipitate   An insoluble solid that has been formed in a liquid by chemical action.

Precipitation   A hazardous waste chemical treatment method by which dissolved material falls out of the waste solution; process is enhanced by the addition of chemicals which induce precipitation.

Precision   The degree to which a measurement is reproducible.

Prehospital Time (EMS)   Interval between activation of the emergency medical transport response to an incident and arrival of the patient at a receiving facility.

Pre-Incident Planning   The process associated with preparing for the response to a hazard by developing plans, identifying resources, conducting exercises, and other techniques to improve an agency's or organization's response capabilities.

Pressure Regulator   Device for maintaining pressure in a line, downstream from the valve.

Pressurized Gases   Gases that are stored in pressure cylinders and remain gases at normal temperature.

Pressurized Water Reactor   A power reactor in which heat is transferred from the core to a heat exchanger by water kept under high pressure to achieve high temperature without boiling in the primary system; steam is generated in a secondary circuit.

Preventative Actions   Directions given by the Incident Commander at an emergency to prevent the problem from increasing.

Prevention Plan   see RISK MANAGEMENT PREVENTION PLAN.

Primacy   The assumption of a state, with the approval of EPA, of the responsibility to administer and enforce a federal program, such as a hazardous waste disposal program.

Primary Assembly Area   An area designated for the assembly of specific groups of people involved in an incident.

Primary Materials   Virgin or new materials used for manufacturing basic products, such as wood pulp, iron ore, silica sand, and bauxite.

Primary Transport (EMS)   Transportation of an emergency patient from the scene of an incident to a receiving facility.

Private Applicator   A person who has been certified by the Dept. of Food and Agriculture or the County Commissioner's Office to apply, use, or supervise the use of any pesticide that is classified for restricted use for purposes of producing any agricultural commodity on property owned or rented by the applicator or the applicator's employer or, if applied without compensation other than trading of personal services between producers of agricultural commodities, on the property of other persons.

Procedures   Preplanned detailed directions for dealing with specific occurrences.

Processing   Technology used to reduce volume or bulk of municipal or residual waste or to convert part or all of such waste materials to off-site reuse. Includes transfer facilities, compost facilities, and resource recovery facilities.

Prodromal Period   The initial period of acute radiation syndrome; period in which the signs and symptoms are evident, generally from one to four days after exposure. In medicine, a prodrome is an early symptom indicating the start of a disease course.

Product   In pesticides, the substance as it is packaged and sold; usually contains an active ingredient plus adjuvants.

Product Substitution   Replacing a hazardous substance in a process with a less hazardous substance.

Products of Combustion   Chemicals, gases, and substances left after a material has entered into the combustion process. Depending on what is burning, those products may be smoke, tar, ash, carbon dioxide, carbon monoxide, toxic gases, and/or condensed steam.

Proficiency Test (EMS)   A practical examination designed to measure the performance of skills.

Project Engineer   An LOP engineer in charge of overseeing the cleanup process.

Projected Dose   The estimated radiation dose that the population-at-risk may potentially receive as a consequence of a nuclear incident.

Prolonged Exposure   More than a brief or one-time contact with a hazardous material such as radioactivity or a pesticide or the residue of that material.

Promptly Available (EMS)   A medically prudent period of time proportionate to the patient's condition and such that any interval of time between arrival of the patient at the facility and the arrival of the health care personnel should not be deleterious to the patient.

Propellent   A liquid in self-pressurized containers that forces the active ingredient from the container.

Proper Shipping Name   The proper shipping name of a commodity or material is the DOT designated name for that commodity or material. (49 CFR, 172.101)

Properties   The characteristics or traits that describe a material.

155

Proportioning Valve   A device used to balance or divide the air supply between the aeration system and the discharge manifold.

Proposition 65   California Safe Drinking Water Act of 1986.

Propulsion Mechanism   The process by which matter or energy is propelled from its position at rest toward another location.

Protectant   A pesticide that is applied before pests are actually found but where they are expected; a preventative.

Protected Area   (1) The area surrounding a hazardous materials incident that is under the control of the responsible agency and from which unauthorized persons are prohibited. (2) In nuclear site incidents, the fenced area surrounding the nuclear steam supply system, turbine-generator, and the Administration and Control Buildings. (3) An area set aside for the preservation of the ecology, such as a wild-life area.

Protection Factor (PF)   A number used to express the relationship between the amount of gamma radiation that would be received by an unprotected person and the amount that would be received by a person in a shelter. Occupants of a shelter with a PF of 40 would be exposed to a dose rate 1/40th (2-1/2%) of the rate to which they would be exposed if unprotected.

Protective Actions   Procedures taken during or after a hazardous materials incident for the protection of the general public from exposures occurring as a consequence of the incident.

Protective Action Guides   Projected general public radiological dose rate or dose commitment criteria which provide guidance to local county and/or state officials for the implementation of protective actions for the protection of the general public from excessive radiation exposure following a nuclear incident.

Protective Barriers   Physical boundaries used to mark the outer edge of an incident area. In radiation incidents, generally used to control access to the area, to limit exposure, and to prevent spread of contamination.

Protective Clothing   Garments and accessories that prevent the contamination of underlying body surfaces or personal clothing.

Protective Equipment or Gear   Any clothes, materials, or devices offering protection from contaminations, whether radioactive or toxic, or from other hazards such as fire scenes present.

Proton   A component of a nucleus, 2000 times more massive than an electron; differs from a neutron by its positive (+1) electrical charge. The atomic number of an atom is equal to the number of protons in its nucleus.

Provisions   Specific emergency OSHA-required equipment and procedures that the employer must provide.

PSH&L   Pressure Settings High and Low; two points for setting of self-closing valves on pipelines.

PSI   Pounds per Square Inch.

PSIA   Pounds per Square Inch Absolute.

PSIG   Pounds per Square Inch Gauge.

PSTN   Pesticide Safety Team Network; regional teams of the National Agricultural Chemical Association designed to assist with pesticide incidents.

Public Hearing/Meeting   A formal means to inform the public about impending federal, state, or local government actions, and to receive their comments.

Public Information Officer   Individual on the command post staff responsible for providing information to the news media and others who need to know; handles inquiries about the incident.

Public Safety Agency   A functional division of a public agency that provides fire fighting, police, medical, or other emergency services.

Public Service Commission   A state body having control of or regulating public utilities.

Publishing Agent   A person authorized by transportation lines to publish tariffs of rates, rules, and regulations for a railroad.

Pull the Pin   (1) Uncouple a RR car (by pulling up the coupling pin). (2) Leave a job, resign, or retire. (3) To cease an operation.

Pulmonary   Pertaining to the lungs.

Pulse   An electrical signal arising from a single event of ionizing radiation.

Pulse Amplifier   A device designed specifically to amplify the intermittent signals of a radiation detection instrument, incorporating appropriate pulse-shaping characteristics.

Pulsed Reactor   A type of research reactor with which repeated, short, intense surges of power and radiation can be produced. The neutron flux during each surge is much higher than could be tolerated during a steady-state operation.

Pulp   Fiber material produced by chemical or mechanical means from fibrous cellulose raw material and from which paper and paperboard are made.

Pump Off To pump so rapidly that the oil level drops below the standing valve on the pump.

Pump-Off Line A pipeline that usually runs from the tank discharge openings to the front of the trailer; most pumps are mounted on the tractor.

Puncture-Resistant Container A leakproof container designed to safely store and/or transport contaminated sharps for proper disposal.

Pusher An extra engine at the rear of a train used to assist a train in climbing a grade.

Purge (Wells) Pumping out well water to remove drilling debris or impurities; also conducted to bring fresh groundwater up into the casing for sample collection. The latter is a means of collecting a representative water sample from the aquifer being investigated.

Purgeable Organic An organic chemical with a high vapor pressure that can be removed from water by bubbling a nonreactive gas such as helium in the water.

PVC Polyvinyl chloride.

PWR see PRESSURIZED WATER REACTOR.

Pyrolysis The process of chemically decomposing an organic substance by heating in an oxygen-deficient atmosphere. High temperatures and closed chambers are used. Major products from pyrolysis of solid waste are water, carbon monoxide, and hydrogen. Some processes produce an oil-like liquid of undetermined chemical composition; gas may contain hydrocarbons and frequently there is process residue of a carbon char. All processes leave a residue of inorganic material. Gaseous products cannot be mixed with natural gas in principal distribution systems unless there is additional chemical processing. Applied to solid waste, pyrolysis has the features of effecting major volume reduction while producing storable fuels.

Pyrophoric A characteristic of those materials which, if ground into fine particles, will spontaneously ignite when exposed to air.

Pyrophoric Liquid Any liquid that ignites spontaneously in dry or moist air at or below 130°F (54°C).

$$Q$$

Q Unit used to express very large energy figures; for example, one Q equals one billion billion BTUs.

QA   Quality assurance.

QC   Quality control.

Qualitative Fit Test   A physical testing of a breathing apparatus face piece to the wearer, performed in an atmosphere of amyl acetate or irritant smoke to evaluate whether the wearer can detect the contaminant, indicating mask leakage and improper fit.

Quality Factor   A number that accounts for the differing ionization potentials of alpha, beta, and gamma radiation; the factor by which absorbed dose is to be multiplied to obtain a quantity that expresses on a common scale the irradiation incurred by exposed persons.

Quantum   Unit quantity of energy according to the Quantum Theory. The photon carries a quantum of electromagnetic energy.

Quantum Theory   The statement by German physicist Max Planck that energy is not emitted or absorbed continuously but in units or quanta; corollary is that the energy of radiation is directly proportional to its frequency.

Quench   To limit or stop the electrical discharge in an ionization detector.

R

R   Rankine temperature scale.

Rabbit   (1) A device to move a sample rapidly from one place (such as inside a research reactor) to another place (such as a radiochemistry laboratory); often are small cylinders of aluminum or plastic, moved by air pressure through a long pipe. (2) Small plug run through an oil flow line by pressure to clean the line or test for obstructions.

RAD   Radiation Absorbed Dose; basic unit of absorbed dose of ionizing radiation; the absorption of 100 ergs of radiation energy per gram of absorbing material.

Radiation   (1) The emitting of energy from an atom in the form of particles or electromagnetic waves; energy waves that travel with the speed of light, and upon arrival at a surface are either absorbed, reflected, or transmitted; not absorbed by air and travel in straight lines; is the most serious cause of fire spreading from one structure to another. (2) Nuclear radiation is that emitted from atomic nuclei in various nuclear reactions.

Radiation Accidents   Accidents resulting in the spread of radioactive material or in the exposure of individuals to radiation.

Radiation Area   Any accessible area in which the level of radiation is such that a major portion of an individual's body could receive in one hour a dose in excess of five millirem or in any five consecutive days a dose in excess of 150 millirem.

Radiation Burn   Radiation damage to the skin; beta burns result from skin contact with or exposure to emitters of beta particles while flash burns result from sudden thermal radiation.

Radiation Chemistry   Branch of research concerned with the chemical effects including decomposition of particles of matter.

Radiation Damage   General term for the harmful effects of radiation on matter.

Radiation Detection Instruments   Devices that detect and record the characteristics of ionizing radiation; *see* COUNTER, DOSIMETER, RADIATION MONITORING.

Radiation Equivalent Man (REM)   Unit of dose equivalent; takes into account the effectiveness of different types of radiation.

Radiation Illness   An acute organic disorder that follows exposure to relatively severe doses of ionizing radiation; characterized by nausea, vomiting, diarrhea, blood cell changes, and, in later stages, hemorrhage and loss of hair.

Radiation Monitoring   Continuous or periodic determination of the amount of radiation present in a given area.

Radiation Protection   Measures to reduce exposure to radiation.

Radiation Protection Guide   The officially determined radiation doses that should not be exceeded; established by the Federal Radiation Council; equivalent to what was formerly termed Maximum Permissible Dose or Exposure.

Radiation Saturation   A phenomenon in which a survey meter's capability to measure radiation levels is overwhelmed, causing the meter to incorrectly read zero.

Radiation Shielding   Reduction of radiation by interposing a shield of absorbing material between any radioactive source and a person or object.

Radiation Source   Usually a man-made, sealed source of radioactivity used in teletherapy, radiography as power source for batteries, or in various types of industrial gauges; machines such as accelerators, radioisotopic generators, and natural radionuclides may also be considered sources.

Radiation Standards   Exposure standards, permissible concentrations, and rules for safe handling; regulations for transportation and for industrial control of radiation and radiation exposure, set by legislative means.

Radiation Sterilization   Use of radiation to cause a plant or animal to become incapable of reproduction; the use of radiation to kill all forms of life, especially bacterias, in food and surgical sutures.

Radiation Therapy   Treatment of disease with any type of radiation; radiotherapy.

Radiation Warning Symbol   A magenta trefoil on a yellow background which is an officially prescribed symbol and should always be displayed when a radiation hazard exists.

Radioactive   Exhibiting or pertaining to radioactivity.

Radioactive Cloud   Mass of air and vapor in the atmosphere carrying radioactive debris.

Radioactive Dating   A technique for measuring the age of an object or sample of material by determining the ratios of various radioisotopes or the products of radioactive decay it contains; example, the ratio of carbon-14 to carbon-12 reveals the approximate age of bones, pieces of wood, or other archeological specimens that contain carbon extracted from the air at the time of their origin.

Radioactive Material (RAM)   Any material that spontaneously emits ionizing radiation.

Radioactive Tracer   A small quantity of radioactive isotope used to follow biological, chemical, or other processes by detection, determination, or localization of the radioactivity.

Radioactive Wastes   Conventional materials that have been contaminated with radiation; not classified as hazardous and not covered by RCRA, they are specifically controlled by the U.S. Atomic Energy Act.

Radioactivity   Spontaneous decay or disintegration of an unstable atomic nucleus accompanied by the emission of radiation.

Radioassay   The analysis of any substance (food, soil, water, etc.) to determine the presence and magnitude of radioactive contamination.

Radiobiology   Body of knowledge and the study of the principles, mechanisms, and effects of ionizing radiation on living matter.

Radio Cache (ICS)   Portable radios, a base station, and repeater stored in predetermined location for dispatch to emergency incidents.

Radio-Controlled Engine (RR)   An unmanned engine situated

161

within the train consist, separated by cars from the lead unit but controlled from it by radio signals.

Radioecology   Study of the effects of radiation on species of plants and animals in natural communities.

Radiography   The use of ionizing radiation for the production of shadow images on a photographic emulsion; some of the rays (gamma rays or X-rays) pass through the subject while others are partially or completely absorbed by the more opaque parts of the subject and thus cast a shadow on the photographic film.

Radioisotope   An unstable isotope of an element that decays or disintegrates spontaneously, emitting radiation. More than 1300 natural and artificial radioisotopes have been identified.

Radiological   Refers to processes involving nuclear radiation.

Radiological Emergency Response Plan   A detailed program of the actions and responsibilities of local, state, or federal government agencies during a radiological emergency.

Radioluminescence   Visible light caused by radiations from radioactive substances; example, the glow from luminous paint.

Radiomutation   A permanent, transmissible change in form, quality, or other characteristic of a cell or offspring from the characteristics of its parent, due to radiation exposure.

Radiopharmaceutical   A material containing radioisotopes used in medical diagnosis or therapy.

Radium   A radioactive metallic element with atomic number 88. An alpha- and gamma-emitter, used as a source of luminescence and as a radiation source in medicine and radiography.

Radius of Influence   The horizontal distance from the center of a well to the limit of the cone of depression.

Radius Rod   A member used to retain axle alignment, and in some cases, control axle torque; can be fixed or adjustable in length.

Rankine Temperature Scale   A scale with the degree interval of the Fahrenheit scale and the zero point at absolute zero; water freezes at 491.60R and boils at 671.69R.

Rapid   Railroad Accident Prevention and Immediate Deployment.

Rate   The amount of pesticide (or pesticide formulation) that is being delivered to a plant, animal, or area; measurement of the volume being applied; expressed in gallons per acre, minute, or hour or other set measure such as pounds per acre, minute, or hour.

Rate of Combustion   see FLAME SPREAD.

RCRA   Resource Conservation and Recovery Act (1976); established a framework for the proper management and disposal of all wastes. RCRA directed EPA to identify hazardous wastes, both generically and by listing specific wastes and industrial process waste streams. Generators and transporters are required to use good management practices and to track the movement of wastes with a manifest system. Owners and operators of treatment, storage, and disposal facilities also must comply with standards, which are generally implemented through permits issued by EPA or authorized states.

Reaction   (1) Chemical: the combining of two or more materials into another, or the chemical breakdown of a chemical material, or the combining of a chemical into a polymer, including oxidation. (2) Physical: the response to a force applied.

Reactive Materials   Substances capable of or tending to react chemically with other substances.

Reactivity   The degree of ability of one substance to undergo a chemical combination with another substance; the tendency to explode under normal management conditions, to react violently when mixed with water, or to generate toxic gases.

Reactor   see NUCLEAR REACTOR.

Reagent   A substance used in chemistry to detect, measure, or produce another substance.

Rear of a Signal (RR)   The side of the signal from which the indication is received.

Receiving Facility (EMS)   A general acute care (hospital) facility assigned a role in the EMS system by the local EMS agency.

Receiving Track   A track used for arriving trains.

Recharge Zone   Area through which water enters an aquifer.

Reclamation   Restoration to a better or more useful state, such as land reclamation by sanitary landfilling or obtaining useful materials from solid waste; use of items for purposes which may be different from their original use.

Recommendation   A suggestion from or advice given by an authority.

Reconsignment   A service extended by a railroad to the owner of the freight permitting a change to the waybill in the name of the shipper, consignee, route, or other instructions to effect delivery of the car providing no back haul is involved.

Recorder   see TECHNICAL SPECIALIST, HAZARDOUS MATERIALS RESPONSE TEAM.

163

Recoverable With solid waste, the capability and likelihood of regaining materials for commercial or industrial use.

Recoverable Resources Materials that still have useful chemical or physical properties after serving a specific purpose and can be reused or recycled for the same or other purposes.

Recovery In oil spills, the process of the physical removal of spilled oil from land, water, or shoreline environments. Methods of oil recovery from water include mechanical skimmers, sorbents, and manual recovery by the cleanup work force; main method of oil recovery spilled on land or shorelines is excavation of contaminated materials.

Recovery Action Procedures taken after the emergency to restore the affected area as nearly as possible to pre-emergency conditions.

Recovery Drum A drum used to overpack.

Recovery Period A period of acute radiation syndrome in which the human body recovers from the effects of acute radiation exposure; follows the period of manifest illness.

Recycle To redirect or utilize a waste or a substance from a waste in a manner that, in the judgment of the department, will not result in a substantial hazard to the health or safety of persons or to livestock, wildlife, or the environment (Section 66174, Title 22, California Administrative Code).

Recycling A resource recovery method involving collection and treatment of waste product for use as a raw material in the manufacture of the same or similar product; also the marketing of products of recycled and recyclable materials.

Reduced Speed Prepared to stop short of another train or obstruction.

Reduction Removal of oxygen from a compound; lowering of oxidation number resulting from a gain of electrons.

Re-Entry Interval Period of time between a pesticide application and when workers can safely go back into an area without wearing protective clothing or equipment.

Reference Library A selection of chemical text books, reference books, microfiche, and computer data programs typically carried by a hazardous materials response team.

References Specific texts or books to consult for information.

Refining in Transit The stopping of shipments of sugar, oil, etc., to be refined at a point located between the points of origin and destination.

Reflector   A layer of material immediately surrounding a reactor core that scatters back or reflects into the core many neutrons that would otherwise escape; returned neutrons can then cause more fission and improve the neutron economy of the reactor; common reflector materials are graphite, beryllium, and heavy water.

Reflex Reflector   Devices used on side or rear of vehicles to give a warning indication to the driver of an approaching vehicle by reflected light from the headlights of the approaching vehicle.

Refrigeration Charge (RR)   A fixed charge for refrigeration from shipping point to destination or for a portion of the trip.

Refrigeration Unit   Cargo space cooling equipment.

Refrigerator Car   A RR car with insulated walls, floor, and roof for carrying commodities that need cooling in transit; two types are those using ice and those using mechanical means.

Regional   An administrative area, division, or district.

Regional Disaster Medical/Health Coordination (RDMHC) (EMS) The individual health officers selected to preplan mutual aid involving mass casualties.

Regional Plan   A hazardous material plan developed pursuant to SARA Title III.

Regional Response Team   Composed of representatives of the federal agencies and a representative from each state in the ten federal EPA regions.

Regional Water Quality Control Boards   In California, organizations from each of the nine regions that formulate and adopt water quality control plans for their respective regions and regulate waste discharges from point and nonpoint sources by establishing and enforcing waste discharge requirements.

Registered Pesticide   A pesticide approved by the U.S. Environmental Protection Agency for use as stated on the label of the container.

Registry of Toxic Effects of Chemicals (RTEC)   Volumes containing more than 58,000 toxicity evaluations of specific chemicals and formulations.

Regular Train   Train authorized by a timetable schedule.

Regulated Pest   A specific organism considered by a state or federal agency to require regulatory restrictions, regulations, or control procedures in order to protect the host, people, and/or the environment.

Regulatory Officials   People who work for the federal or state government and enforce rules, regulations, and laws.

Re-Kindle   The process in which a controlled or apparently extinguished fire returns to the free-burning stage; generally caused by improper extinguishment practices and lack of overhaul.

Relative Biological Effectiveness (RBE)   Factor used to compare the biological effectiveness of different types of ionizing radiation; inverse ratio of the amount of absorbed radiation required to produce a given effect to a standard radiation required to produce the same effect.

Relative Humidity   The percentage of moisture in given volume of air at a given temperature in relation to the amount of moisture the same volume of air would contain at the saturation point.

Relay Emergency Valve   A combination valve in an air brake system that controls brake application and also provides for automatic emergency brake application should the trailer become disconnected from the towing vehicle.

Release   Escape of radioactive materials into the noncontrolled environment.

Release Cock   *see* RELEASE VALVE.

Release Rod (RR)   A small iron rod generally located at the side of a car for the purpose of operating the air brake release valve.

Release Valve (RR air brake)   A valve attached to the auxiliary reservoir for reducing the air pressure when a locomotive is detached so as to release the brakes.

Relief Valve   A valve that will open automatically when pressure gets too high.

REM   *see* RADIATION EQUIVALENT MAN.

Remedial Action   Action taken to correct a problem such as fuel contamination of soil and groundwater.

Remote Assembly Area   An area designated outside a hazardous materials incident for the assembly of personnel.

Remote Control Station   A station containing equipment to control and regulate operations in an oil field.

Remote Unit   *see* RADIO-CONTROLLED ENGINE.

Remote Sensing   Aerial sensing of oil on a water surface; primary applications are the location of an oil spill prior to its detection by any other means and monitoring of the movement of an oil slick under adverse climatic conditions and during the night.

Removal   Frequency and immediacy of clothing removed in case of contamination; information in compliance with government standards or recommendations from recognized authorities.

Removal Action   *see* MITIGATION.

Repair Track   RR track used for repair of cars.

Repeated Contact or Inhalation   Exposure to or breathing in of a pesticide on several occasions over a period of time.

Repellent   A pesticide that makes pests leave or avoid a treated area, surface, animal, or plant.

Reportable Incident   Any incident that has or may impact the public health safety or the environment, or is otherwise required by law to be reported.

Reportable Quantity (RO)   The designated amount of a specific material that if spilled or released requires immediate notification to the National Response Center (NRC) (see 49 CFR 172.101, 40 CFR 117.3, 173, and 302.6).

Representative Sample   A sample that is assumed not to be significantly different than the population of samples available. In fuel leak investigation, samples are often selected to be representative of the worst case situation.

Rescue   The removal of victims from an area determined to be contaminated or otherwise hazardous by appropriately trained and equipped personnel.

Residence Time   The time during which radioactive material remains in the atmosphere following the detonation of nuclear explosive; usually expressed as a half-time, since the time for all material to leave the atmosphere is not well known; half-time.

Residential Waste   Waste materials generated in homes and apartments, including paper, cardboard, beverage containers, food cans, plastics, food wastes, glass, garden, and yard wastes.

Residual Nuclear Radiation   Radiation emitted by radioactive material remaining after a nuclear explosion; arbitrarily designated as that emitted more than one minute after the explosion.

Residual Oils   The oil remaining after fractional distillation during petroleum refining; generally includes the bunker fuel oils.

Residual Pesticide   A pesticide remaining in the environment for a fairly long time; may continue to be effective for days, weeks, and months.

Residuals Repository   A conceptual hazardous waste disposal facility specifically restricted to receiving only residuals from hazardous waste treatment facilities. Potentially, wastes could be kept segregated to allow eventual reclamation.

167

Residual Waste   Garbage, refuse, or other waste including solid, industrial, mining, and agricultural operations and sewage from an industrial, mining, or agricultural water supply treatment facility, waste water treatment facility, or air pollution control facility, provided it is not hazardous.

Residue   A hazardous material remaining in a packaging after its contents have been emptied and before the packing is refilled or cleaned and purged of vapor to remove any potential hazard.

Residue Tolerance   The amount of pesticide residue permissible in or on a raw agricultural commodity.

Resistant   Not affected by a potentially injurious intruder; when an organism is not affected by a pesticide in the manner or to the degree that was expected or that another organism was affected; tolerant.

Resistant Species   A species difficult to kill or control with a particular pesticide; one able to suppress or retard the damage usually done by a pesticide.

Resources   All of the immediate or supportive assistance available to help control an incident, including equipment, material, personnel, control agents, agencies, and printed emergency guides.

Resource Conservation   Reduction of the amounts of solid waste that are generated; reduction of overall resource consumption and utilization of recovered resources.

Resource Conservation and Recovery Act (RCRA)   see RCRA.

Resource Recovery   The salvage of discarded hazardous materials or their conversion into a reusable, saleable, or valuable form. Salvaged or converted materials shall not be considered waste (Section 66180, Title 22, California Administrative Code; the Department of Health Services is proposing to revise this definition).

Respirator   A face mask that filters out poisonous gases and particles from the air, enabling a person to breathe and work safely; used to protect the nose, mouth, and lungs from hazardous materials.

Respiratory   Having to do with breathing (inhalation); of the lungs, nose, mouth, and oxygen supply to a human or animal.

Respiratory Protective Equipment   see SELF-CONTAINED BREATHING APPARATUS.

Respiratory Toxicity   How poisonous a pesticide is to an animal or person when breathed in through the lungs; an intake of any toxic substance through air passages into the lungs.

Response That portion of incident management in which personnel are involved in controlling a hazardous material incident. (NFPA 472, 1–3)

Response Time Interval of time from initiation of an alarm requiring emergency response vehicle activation to the vehicle's arrival at the scene of the incident.

Response Trust Fund A $1.6 billion fund used for cleanup of abandoned and existing disposal sites; sources of this fund are industrial taxes on oil and certain chemical feedstocks (87%) and federal appropriations (13%).

Responsible Party (RP) A legally recognized entity (person, corporation, business, partnership, etc.) that has a legally recognized status of financial accountability and liability for action necessary to abate and mitigate adverse environmental and human health and safety impacts resulting from a nonpermitted release or discharge of hazardous material. The person or agency found legally accountable for the cleanup of the incident.

RG Registered Geologist.

Restricted Speed (RR) Proceed prepared to stop short of train; obstruction, or switch not properly lined, not exceeding 15 mph.

Restricted Use Pesticide A pesticide that has been classified under provision of FIFRA (amended) for use only by an appropriately certified applicator; a limited use pesticide that is applied only by qualified personnel according to the law.

Restrictions Limitations.

Retaining Valve A small, manually positioned valve located near the brake wheel for retaining part of the brake cylinder pressure to aid in retarding the acceleration of a train in descending long grades.

Retarder (RR) A metal grip adjacent to the rails, usually operated by compressed air or electric motors, for regulating speed of a car by pressure on the wheels while rolling down a hump incline.

Retention Time The time hazardous waste is subjected to the combustion zone temperature in an incinerator.

Reuse Multiple use of a product in its original form; differs from recycling in that the product is not reprocessed or refabricated before it is used again.

Revenue Waybill A document showing the amount of charges due on a shipment.

Reverse Lever (RR) The lever that controls the direction of motion of a locomotive by reversing the traction motor field connections.

169

Richter Scale   A measure of the amplitude of earthquake waves at their point of origin, giving an indication of how much energy has been released.

Right-of-Way   Property owned by a railroad over which tracks have been laid.

Ringer Solution   A water solution containing chemical constituents simulating the blood of animals in which tissues can be kept in a living state for various periods of time.

Ring Stiffener   A circumferential tank shell stiffener which helps to maintain the tank cross section.

Rip   RR car in need of repair.

Rip Track   *see* REPAIR TRACK.

Riser   A pipe through which liquid travels upward.

Risk   (1) The combination of probability of an accidental occurrence and the likely magnitude of consequences in a given exposure. (2) A situation which may result in death or injury to persons or in damage to property; including effects of toxicity, fire, explosion, shock, concussion, fragmentation, and corrosion.

Risk Analysis   A process to analyze the probability that harm may occur to life, property, and the environment and to note the risks to be taken in order to identify the incident objectives.

Risk Assessment   Evaluation of the threat to public health and the environment posed by a hazardous waste facility; considering the probability of an incident and its effects.

Risk Management   Decision-making process that involves such considerations as risk assessment, technological feasibility, economic information about costs and benefits, statutory requirements, public concerns, and other factors.

Risk Management Prevention Plan (RMPP)   Statutory requirements in Health and Safety Code, Section 25534, subsection (1). A plan that encompasses, among other appropriate elements:

- a structured assessment of hazards
- a formal personnel training program for the prevention of, and response to, emergencies
- procedures for periodic safety reviews of operating equipment and procedures
- schedules for regular testing of the program and
- procedures for the purpose of reducing the probability of accidents.

RNA (Ribonucleic acid)   An essential component of all living matter; one form carries genetic information.

Roadbed   Foundation on which the railroad track and ballast rest.

Road Haul (RR)   Line haul.

Roadside   Side of a trailer farthest from the curb when trailer is traveling in a normal forward direction (left-hand side); opposite of curbside.

Roadway   *see* RIGHT-OF-WAY.

Rodent   Any animal of the order Rodentia; mice, rats, squirrels, gophers, woodchucks.

Rodenticide   A pesticide used to control rodents.

Roentgen (R)   The unit of radiation exposure in the air; units for quantities of X-ray or gamma radiation measured by detection and survey meters. Named after Wilhelm Roentgen, German scientist who discovered X-rays in 1895.

Roll-Off Truck   A vehicle that deposits and collects a ten- to fifty-cubic yard container at a site; generally employed in industrial waste collection systems.

Roller Bearings   General term applied to a group of journal bearings which depend upon the action of a set of rollers to reduce rotational friction.

Rotary Gauge   A variable liquid gauge consisting of a small positive shutoff valve located at the outer end of a tube, the bent inner end resting in the container interior; tube is installed so the tube can be rotated with a pointer on the outside to indicate the relative position of the bent inlet end. The length of the tube and the configuration to which it is bent is suitable for the range of liquid levels to be gauged. The level in the container where the inner end begins to receive liquid can be determined by the pointer position on the outside scale at which a liquid-vapor mixture is observed to be discharged from the valve.

Roughing Filter   A prefilter with low efficiency for small particles, usually of the panel type.

Route   (1) The course or direction that a shipment moves. (2) To designate the course or direction a shipment shall move.

Rover   A joint program of the Atomic Energy Commission and the National Aeronautics and Space Administration to develop a nuclear rocket for space flight.

RPAR   Rebuttable Presumption Against Registration; an EPA process to identify pesticide chemicals presenting "unreasonable adverse effects on the environment."

RPM Revolutions per minute; a measurement of the speed at which something turns or spins.

RRT Regional Response Teams; composed of representatives of federal agencies and a representative from each state in the federal region. During a response to a major hazardous materials incident involving transportation or a fixed facility, the OSC may require that the RRT be convened to provide advice or recommendations in specific issues requiring resolution. Under the NCP, RRTs may be convened by the chairman when a hazardous materials discharge or release exceeds the response capability available to the OSC in the place where it occurs; crosses regional boundaries; or may pose a substantial threat to the public health, welfare or environment or to regionally significant amounts of property. Regional contingency plans specify detailed criteria for activation of RRTs. RRTs may review plans developed in compliance with SARA Title III, if the local emergency planning committee so requests.

RTEC see REGISTRY OF TOXIC EFFECTS OF CHEMICALS.

RTEC Number Number assigned by NIOSH to facilitate quick location in the Registry of Toxic Effects of Chemicals.

Runaway Pests Any pest organisms entering a new territory where they have no natural enemies and therefore reproduce with little interference, resulting in a large population which can overrun an area.

Rubbish A term for solid waste that does not include food wastes.

Rule "G" A railroad operating rule prohibiting the use or possession of intoxicants or narcotics while on duty.

Running Gear A general term applied to and including the wheels, axles, springs, axle boxes, frames, and other carrying parts of a truck or locomotive.

Running Lights Vehicle marker, clearance, and identification lights required by regulations.

Running Track (1) A railroad track designated in the timetable upon which movements may be made subject to prescribed signals and rules or special instructions. (2) A track reserved for movement through a yard.

Runoff (1) Rainwater, leachate, or other liquid that drains off the surface of the ground. (2) A sprayed liquid (pesticide) which does not remain on the surface of a plant.

Runon Rainwater, leachate, or other liquid that drains overland onto part of a facility.

Runway   Defined area prepared for landing and takeoff run of aircraft along its length.

Rupture   The physical failure of a container or mechanical device, releasing or threatening to release a hazardous material (Sacramento Fire Department, HMRT).

Rupture Disc   A safety relief device in the form of a metal disc that closes the relief channel under normal conditions; disc bursts at a set pressure to permit the escape of gas.

RVR   Runway Visual Range; maximum distance in the direction of takeoff or landing at which the runway or the specific lights or markers delineating it can be seen from a position above a specified point on its center line at a height corresponding to the average eye level of pilots at touchdown (approximately 16 feet).

RWQCB (or Regional Board)   see REGIONAL WATER QUALITY CONTROL BOARD.

S

SAE   Society of Automotive Engineers; a professional organization dedicated to the advancement of the knowledge, experience, and skill in automotive engineering.

Safener   A chemical added to a pesticide to keep it from injuring plants.

Safety Chain   A secondary connection consisting of chains or cables used between the towing and towed vehicles to retain emergency hook-up in event of separation of the primary towing connection.

Safety Officer   Individual on the command post staff responsible for the well-being of those operating at a hazardous materials incident; position requires cooperation with many different agencies.

Safety Relief Valve   Device on pressure cargo tanks containing an operating part held in place by spring force; valve opens at set pressures.

Safety Rod   A standby control rod used to shut down a nuclear reactor rapidly in an emergency.

Salivation   An excessive discharge of saliva; ptyalism.

Salt   Compound of the negatively charged ion from an acid and the positively charged ion from a metal or alkali base.

173

Salvage   The extraction or controlled removal of materials from the solid waste stream for reuse.

Salvage Cover   A large vinyl or canvas cover used to protect building contents from water during firefighting operations.

Salvage Drum   *see* RECOVERY DRUM.

Sample   Material containing radionuclides; may consist purely of radionuclides or be a mixture of radionuclides and non-radioactive material.

Sampling Kits   Kits assembled for the purpose of providing adequate tools and equipment for the taking of samples and documentations of unknowns to create a "chain of evidence."

Sanded Up   Clogged by sand entering a well bore with the oil.

Sanders   Devices operated by air for applying sand to the rail in front of or behind the driving wheels of the railroad engine.

Sandshoe   A flat steel plate serving as ground contact on the supports of a trailer and used instead of wheels, particularly where the ground surface is expected to be soft.

Sanitary Landfill   A method of disposing of refuse, usually below ground level, without creating nuisances or hazards to public health and safety.

Sanitary Landfill Site   An approved disposal area where materials including garbage, oil-contaminated debris, and other solid wastes are spread in layers and covered with soil to a depth that will prevent disturbance or leaching of contaminants toward the surface. Careful preparation of the fill area and control of water drainage are required to assure proper landfilling. Refuse is confined to the smallest practical area and reduced to the smallest practical volume by using heavy tractor-like equipment to spread, compact, and usually cover the waste daily with at least six inches of compacted soil. After the area has been completely filled and covered with a final two- to three-foot layer of soil and allowed to settle an appropriate period of time, the reclaimed land may be utilized as a recreational area such as a park or golf course; under certain highly controlled conditions the land may be used for some types of building constructions.

SARA Title II Regional Plan   see REGIONAL PLAN.

Saturated Zone   The part of the earth's crust in which virtually all voids are filled with water.

Scale House   Structure housing weight-recording mechanism used in weighing freight cars.

Scale Test Car   A compact RR car equipped with weights for the testing of track scales.

Scale Track   A storage track for RR cars needing to be weighed.

Scaler   An electronic instrument for rapid counting of radiation-induced pulses from Geiger counters or other radiation detectors; permits rapid counting by reducing (by a definite scaling factor) the number of pulses entering the counter.

Scanner   Device used to determine location and amount of radioactive isotopes within the body by measurements taken with instruments outside the body; moves in a regular pattern over area to be studied or over the whole body and makes a visual record.

Scattering   A process changing a particle's trajectory, caused by particle collisions with atoms, nuclei, and other particles or by interactions with fields of magnetic force; see COMPTON EFFECT.

Scavenging   (1) In chemistry, the use of a nonspecific precipitate to remove one or more undesirable radionuclides from solution by absorption or coprecipitation. (2) In atmospheric physics, the removal of radionuclides from the atmosphere by action of rain, snow, or dew; see FALLOUT.

SCBA   see SELF-CONTAINED BREATHING APPARATUS.

Scenario   An outline of a natural or expected course of events.

Scene   The location impacted or potentially impacted by a hazard.

Scene Manager   see INCIDENT COMMANDER.

Schedule   Part of a timetable that prescribes class, direction, number, and movement for a regular train.

Scintillation   A flash of light produced in a phosphor by an ionizing event. Compare with fluorescence, luminescence.

Scintillation Counter   An instrument that detects and measures ionizing radiation by counting the light flashes caused by radiation impinging on certain materials (phosphors).

Scram   The sudden shutdown of a nuclear reactor, usually by rapid insertion of the safety rods, caused by an emergency or deviation from normal reactor operation.

Scrap   Waste material suitable for recovery or reclamation.

Scraper   Device used to clean deposits of paraffin from tubing or flow lines (in oil industry).

175

Screen   Perforations in a well casing and usually located near the bottom of the well or at selected depths to tap perched aquifers.

Scrubber   Device using a liquid filter to remove gaseous and liquid pollutants from air stream.

SDOHS   State Department of Health Services.

SDV   Shut Down Valve; A valve put into a pipeline to shut the system down if the pressure in the system goes over a certain pressure or drops below a certain pressure.

Seals   Metal strips designed for one-time use applied to hasp of closed freight containers; must be broken to remove and are used to indicate that container contents have been undisturbed while in transit; stamped with a name, initial, and/or number for identification.

Secondary Materials   All types of materials handled by dealers and brokers that have fulfilled their useful function and usually cannot be used further in their present form or at present location, and materials that occur as waste from the manufacturing or conversion of products.

Secondary Track   A designated track where trains may be operated without timetable authority, train orders, or block signals.

Secure Landfill   Landfill constructed in cell forms to segregate and isolate hazardous materials from each other; clay, plastics, pozzolanic liners form separated, permanent entrapments.

Sediment   Material in suspension in air or water; the total dissolved and suspended material transported by a stream or river.

Sedimentation   Hazardous waste physical treatment method that separates and removes by gravitational settling suspended particles that are heavier than the liquid in which they are present.

Seebeck Effect   Phenomenon involved in operation of a thermocouple; named for German scientist Thomas Seebeck, who first observed concept in 1822.

Seed Protectant   A chemical applied to seed before planting to protect seeds and new seedlings from diseases and insects.

Selective Pesticide   A pesticide more toxic to some types of plants or animals than to others; a selective herbicide might kill crabgrass in a cornfield but would not injure the corn.

Selective Toxicity   The capacity of a chemical to injure one kind of living matter without harming another, even though the two may be in intimate contact.

Self-Accelerating Decomposition Temperature   The temperature

176

above which the decomposition of an unstable material proceeds by itself independently of the external temperature.

Self-Aligning Coupler (RR)   A coupler having a taper shank rather than a straight shank to prevent the jackknifing of cars.

Self-Contained Breathing Apparatus (SCBA)   A positive-pressure, self-contained breathing apparatus or combination SCBA/supplied-air breathing apparatus certified by the National Institute for Occupational Safety and Health (NIOSH) and the Mine Safety and Health Administration (MSHA) or the appropriate approval for use in atmospheres that are immediately dangerous to life or health (IDLH) (NFPA 1991, 1–3).

Semipermeable   Can be penetrated by some gases or liquids and not by others.

Semitrailer   A truck trailer equipped with one or more axles and constructed so the front end and a substantial part of its own weight and that of its load rest upon a truck tractor.

Sensitive   Easily injured or affected by; susceptible to effect with low exposure or dosage.

Sensitive Areas   Locations where pesticide applications could cause great harm; examples: streams, ponds, houses, barns, parks, etc.

Sensitivity Maps   Maps used by an on-scene commander and hazardous materials incident response teams which designate areas of biological, social, and economic importance in a given region; often prioritize sensitive areas; usually contain other information useful to the response team such as access areas, roads, etc.

Sensitizer   A substance which on first exposure causes little or no reaction in man or test animals, but which on repeated exposure may cause a marked response not necessarily limited to the contact site.

Septage   Water waste extracted from septic tanks; disposal is the responsibility of State Water Resources Control Board.

Septicemia   A disease condition commonly known as blood poisoning, caused by harmful bacteria, similar organisms, or their toxic products.

Sesoil   A computer model for predicting the movement/transport of a chemical in the vadose zone.

Set-Up   An article or area put together in its complete state; not knocked down.

Sewage   Waste flushed from toilets into sewers, then processed in treatment plants. Sewage disposal is responsibility of the State

Water Resources Control Board; disposal of sludge or the dried end-product of the sewage treatment process is the responsibility of the State Solid Waste Management Board.

Sharp   Any object that can penetrate the skin including, but not limited to, needles, lancets, scalpels, and broken capillary tubes. Used in describing communicable disease exposure.

Sheer Section   A safety feature in cargo tank piping and fittings designed to fail or break completely to prevent damage to shutoff valves or the tank itself.

Sheet Breakaway   In oil spills, a type of current-induced boom failure resulting from the fact that a boom placed in moving water acts like a dam; surface water being held back by the boom is diverted downwards and accelerates in attempt to keep up with the water flowing under the boom skirt, thus drawing oil from the surface under the boom.

Shelter   A structure or other location offering protection from the elements; in nuclear incidents, a barrier from radiation in the environment.

Shielding   Material that can be used to absorb and block radiation energy.

Shift Technical Advisor   At a nuclear facility, the individual on duty in the control room responsible for advising of plant and systems status.

Shifting (RR)   Switching.

Shipment   Cargo being transported, includes material, its packaging or container, marks, etc.

Shipper   Person or firm originating shipment of goods; consignor.

Shipper's Export Declaration   Form required by Treasury Department, prepared by consignor showing value, weight, consignee, destination, etc., of shipments to be exported.

Shipper's Load and Count   Verification by the consignor that car contents were loaded and counted by the shipper and not the railroad.

Shipping Document   A shipping order, bill of lading, manifest, or other document issued by the transport carrier.

Shipping Order, Papers   Instructions to the railroad or transport carrier for forwarding all goods; usually the second copy of the bill of lading.

Shock   Severe reaction of the human body to a serious injury; can result in death if not treated, even if actual injury was not fatal.

Shock Wave A pressure pulse in air, water, or earth, propagated from an explosion which has two phases; in the first (positive) the pressure rises sharply to a peak then subsides to the normal pressure of the surrounding medium; in the second (negative) the pressure falls below that of the medium then returns to normal.

Shop (RR) Structure where building of and repair to equipment is performed.

Shoreline Sensitivity The susceptibility of a shoreline environment to any disturbance that might decrease its stability or result in short- or long-term adverse impacts; shorelines most susceptible to damage are usually equally sensitive to cleanup activities which may alter physical habitat or disturb associated flora and fauna.

Shoreline Type The average slope or steepness and predominant substrate composition of a shoreline area.

Short-Term Exposure Limit (STEL) A fifteen-minute time-weighted average exposure that should not be exceeded at any time during a work day, nor repeated more than four times per day, even if the eight-hour time-weighted average is within the Threshold Limit Value (TLV).

Short-Term Pesticide A pesticide that breaks down into nontoxic by-products almost immediately after application.

Short Ton 2,000 pounds.

Shredder Mechanical device used to break up waste materials into smaller pieces by tearing and impact actions.

SI The International System of Units; adopted by the General Conference on Weights and Measures, an international diplomatic conference dedicated to developing and unifying the metric system.

SIC Standard Industrial Code; prepared by the United States Office of Management and Budget.

SIC Number A number assigned to a corresponding type of industry, manufacture, or product under the Standard Industrial Code.

Sidedress Pesticide application along the side of a crop row.

Side Loader A refuse truck in which solid waste is loaded into the side of the vehicle.

Side Marker Lamps Lamps that show to the side of a vehicle; one red lamp mounted as far to the rear as practical, and one amber lamp mounted as far to the front as practical, not less than fifteen inches above the road surface to show the extreme overall length of the vehicle; an intermediate marker lamp is required on trailers over thirty feet in length, mounted midway between the forward and aft marker lamps.

Side Rails   The main longitudinal frame members of a tank container used to connect the upper and lower corner fittings.

Side-Track   A RR track adjacent to the main track for purposes other than meeting and passing trains.

Siding   A RR track adjacent to a main or a secondary track for meeting or passing trains.

Sievert (Sv)   The SI unit for measuring dose equivalent; 1 Sv = 1 joule/kg = 100 rems.

Sign   (1) Some evidence of exposure to a dangerous pesticide; an outward signal of a disease or poisoning in a plant, animal, or human. (2) A placard or written display used to inform. (3) A hand gesture.

Signal   Placard, gestures or lights that convey a message.

Signal Words and Symbol   Words that must appear on pesticide labels in accordance with FIFRA (amended) to show toxicity of that pesticide.

Signal Words	Toxicity	Symbol
Caution	Low	
Warning	Moderate	
Danger—Poison	High	Skull & Cross Bones

Signboard (RR)   Information stencilled on car side pertaining to empty car movement instructions.

Sill   The main longitudinal members of a car or truck underframe.

Silt   Soil or sediment particles that range in size from four to sixty-four microns; larger than clays (four microns) and smaller than sand (sixty-four microns to two mm).

Silvicide   A pesticide used to control unwanted brush and trees.

Silviculture   Care and cultivation of forest trees.

Simulation   A mock accident or exercise set up to test emergency response methods or for use as a training tool.

Single Resources (ICS)   Any resource or grouping of resources on which an individual status is maintained.

Single Track   A main RR track upon which trains are operated in both directions.

Sinking Agent   Materials spread over the surface of an oil slick to absorb oil and cause it to sink; rarely used because sinking oil may cause considerable damage to bottom-dwelling organisms; agents include treated sand, fly ash and special types of clay. Provide a purely cosmetic approach to oil-spill cleanup.

SIS   Standardiseringskommissionen i Sverige (Sweden).

Site   (1) A specific location. (2) An area, location, building, structure, plant, animal, or other organism to be treated with a pesticide to protect it from or to reach and control the target pest.

Site Area Emergency   An event involving actual or likely major failures of nuclear plant functions needed for the protection of the public; offsite protective actions unlikely.

Site Assessment   Activities taken to determine the extent of contamination.

Situation Unit Leader (ICS)   Individual responsible for collection and display of status resources.

Skate (RR)   A metal skid placed on rail in a hump yard to stop cars from running out the lower end of the classification yard.

Skew Bridge   A bridge crossing a passageway at other than a right angle.

Skid Tank   Portable tank, usually positioned on skids or a trailer.

Skimmer   Physical systems whereby a liquid phase is recovered from another liquid phase due to polarity differences and stored or transferred for further processing. Typical use is to remove petroleum products floating on a water body.

Skirt   Portion of a floating boom that lies below the water surface and provides the basic barrier to the spread of an oil slick or the loss of oil beneath the boom.

Slack Adjuster   Adjustable mechanical lever used to transmit brake chamber force to brake cam shaft when the brakes are applied; are designed so that they can be adjusted to compensate for lining wear.

Slave Unit   (1) A mechanical device directly responsive to another device. (2) A radio-controlled RR engine.

Slick   Film of oil on water surface (usually less than two microns thick).

Sliding Fifth Wheel   A fifth wheel assembly capable of being moved forward or backward on a truck tractor to vary load distribution on the tractor and to adjust the overall length of combination.

Slimicide   A pesticide used to prevent slimy growths.

Sling   A wire-rope loop for use in lifting heavy equipment.

Slips Wedge-shaped toothed pieces of metal that fit inside a bowl and are used to support tubing or other pipe.

Slip Tube Gauge A variable liquid level gauge with a relatively small positive shutoff valve located at the outside end of a straight tube, normally installed vertically, which communicates with the container interior; tube installation fitting is designed so tube can be slipped in and out of container and the liquid level at the inner end determined by observing when the shutoff valve vents a liquid-vapor mixture.

Slope Sheet Panels located at each end of payload compartment of transport vehicle which direct product by gravity to hoppers.

Slow Board A RR signal indication to proceed at a slow speed.

Slow Speed Not exceeding fifteen miles per hour.

Sludge A heavy, slimy solid, semisolid, or liquid waste generated from a municipal, commercial, or industrial waste treatment facility, wastewater treatment plant, waste supply treatment plant, or air pollution control facility, exclusive of treated effluent from a wastewater treatment plant.

Slurry Thin mixture of liquid (usually water) and fine particles.

Smoke The dense, opaque, gas-like product of a fire; in the chemical sense does not have a fixed composition but consists of many toxic gases, carbon particles, and unburned fuel vapors.

Small Quantity Generators Facilities generating or accumulating less than 1,000 kg/month of identified hazardous waste or specific quantities of acutely hazardous wastes; some are exempted from RCRA.

Smear A wipe sample taken of a particular area to determine the presence of loose radioactive contamination.

SMYS Specified Materials Yield Strength; an indication of a material's ability to withstand pressure.

Snake Out To pull out.

SNAP Systems for Nuclear Auxiliary Power; an Atomic Energy Commission program to develop small auxiliary power sources for specialized space, land, and sea uses.

SNV Swiss Association for Standardization (Switzerland).

SOAR Save Oil America, Recycle; an oil-recycling program directed by the California State Solid Waste Management Board (SSWMB). Program initiated in 1979 as a result of SB68, the Used Oil Recycling Act, passed by the state legislature in 1977. More than 2,400 gas stations, stores, and recycling centers serve as SOAR collection points where "do-it-yourself" oil changers and others can bring their used motor and industrial oil for recycling. Californians are currently recycling almost half of the used oil generated in the state through the SOAR project.

Soil Application   Putting a pesticide in or on the soil instead of applying it directly to vegetables.

Soil Contamination   Contamination of the ground area where a pesticide spill or fire occurs, or where contaminated runoff water flows.

Soil Fumigant   Pesticide used to control pests in the soil; when added to the soil, takes form of gas or vapor; evaporates quickly so often used with some kind of cover (a plastic sheet spread over area) to trap vapor in soil until pest is under control.

Soil Incorporation   The mixing of pesticide into soil by mechanical means, usually consisting of pesticide application to the soil followed by some sort of tillage.

Soil Injection   Mechanical placement of pesticide below ground level into the soil; usually a pesticide liquid which vaporizes.

Soil Sterilant   Chemical prevents growth of all plants and animals in the soil; depending on chemical, may be temporary or long-term.

Sole Use Vehicle   A vehicle designated exclusively for transporting radioactive material.

Solid Track (RR)   Track full of railroad cars.

Solidification   Process of stabilizing waste into a solid with high structural integrity; solidified wastes are less likely to leach out of land disposal sites than are untreated wastes even though the physical and chemical characteristics of the constituents of the waste may not be changed by the process.

Solid Waste   Discarded material other than liquid waste, including paper, dirt, wood, lawn and garden waste, food waste, agricultural waste, industrial waste, demolition and construction waste, glass, ceramics, metal, manure, ash, clothing, rags, leather, rubber, and plastic, resulting from industrial, commercial, mining, agricultural operations, and from community activities and containing resources that can be recycled, reused, or otherwise recovered.

Solid Waste Disposal Act Amendments of 1980   Amends the federal Resource Conservation and Recovery Act of 1976 (RCRA), with established effective solid waste management as a national priority. Concerning nonhazardous solid wastes, the amendments establish minimum federal funding of state and local agencies for developing solid waste management plans, authorize additional funding for rural community assistance, and authorize state plans to address energy and materials conservation and recovery. The Solid Waste Disposal Act reauthorizes state surveys of solid waste disposal sites; technological and financial studies of resource-recovery systems by the Environmental Protection Agency; and development of markets for resource-recovery products by the Secretary of Commerce, as mandated by RCRA. The Used Oil Recycling Act (PL96-483) amended the Solid Waste Disposal Act amendments to relax labeling restrictions on rerefined oil products and provide funding to the states for used oil recycling programs.

Solid Waste Management   The overseeing and regulation of the safe and sanitary reuse or disposal of industrial, residential, construction and agricultural wastes; includes collection, transportation, storage, recycling, resource-reuse, recovery, and waste-reduction programs.

Solubility   The ability of a substance to mix with water; important in firefighting since flash points and fire points of a flammable liquid are appreciably affected as the amount of water in a mixture is increased.

Soluble   Will dissolve in a liquid.

Soluble Powder (SP)   Dry preparation containing a fairly high concentration (15%–95%) of active ingredient that dissolves in water or another liquid to form a solution that may be applied.

Solute   The substance that is dissolved in the solvent to form a solution.

Solution   Mixture of one or more substances into another substance (usually a liquid) in which all ingredients are completely dissolved without their chemical characteristics changing; will not settle out or separate in normal use.

Solvent   Liquid that will dissolve a substance to form a solution; example: water is the solvent when it dissolves sugar. Used in a number of manufacturing/industrial processes including manufacture of paints and coatings for industrial and household purposes, equipment cleanup, and surface degreasing in metal fabricating industries.

SONGS-1   San Onofre Nuclear Generating Station, Unit 1.

Sorbent   A substance capable of either adsorbing or absorbing another substance; a natural organic, mineral-based or synthetic organic material used to recover small amounts of spilled material (such as oil).

Sorbent Barrier   In oil spills, a boom constructed of or including sorbent materials to simultaneously recover spilled oil during the containment process; used only when the oil slick is relatively thin, since recovery efficiency rapidly decreases once the sorbent is saturated with oil.

Sorbent-Surface Skimmer   Mechanical skimmer incorporating a rotating sorbent surface (oleophilic) drum, disc, belt, or rope to which oil adheres as the surface is moved continuously through the slick.

Sour Gas   Gas that contains hydrogen sulphide.

Source   Material that is emitting radiation.

Source Separation   The setting aside of recyclable wastes at their point of origin for purposes of recycling; example: home separation of paper, glass, and metal from other wastes for separate collection by a recycler.

Source Term   A particular type or amount of radionuclide originating at the source of a nuclear incident; also describes conditions and mode of emission.

Space Spray   A pesticide applied as tiny droplets which fill the air and destroy insects and other pests, either indoors or out-of-doors.

Special Equipment (RR)   Freight cars designed to carry specific commodities, some of which contain devices to protect and/or aid in handling shipments.

Special Groups   Concentrations of people in one area or building for special circumstances (e.g., schools, hospitals, nursing homes, shopping centers).

Special Information   Important data not ordinarily found, such as intricate air measuring ratios or new blood level test criteria.

Special Nuclear Material   In atomic energy law, referring to plutonium 239, uranium 233, uranium containing more than the natural abundance of uranium 235, or any material artificially enriched in any of these substances.

Special Protection   A means of limiting the temperature of an LP-Gas container for purposes of minimizing the possibility of failure of the container as a result of fire exposure; may consist of applied insulating coatings, mounding, burial, water spray fixed systems, or fixed monitor nozzles.

Specialized Treatment Facility   A facility designed to treat a specific type or group of hazardous wastes, usually generated by a small number of industries with similar production processes.

Species   Group of living organisms called by the same common name because they are very much alike and can interbreed successfully.

Specific Activity The rate of decay of a radioactive sample per unit of sample weight; measured in curies per gram. Levels of activity and specific activity are measured with regard to the total weight of radioactive and nonradioactive material in the sample.

Specific Category A pesticide use designated by FIFRA or by state regulations; a special area (such as forest pest control or aquatic pest control) requiring state certification for use of restricted materials by commercial applicators.

Specific Gravity Ratio of the weight of the volume of liquid or solid to the weight of an equal volume of water; specific gravity of water is 1.0; material with a specific gravity less than 1.0 will float on water, that with a specific gravity of more than 1.0 will sink in water.

Specific Heat Measurement of the ability of a substance to absorb heat, indicated by the relative quantity of heat needed to raise the temperature of one pound of the substance one degree Fahrenheit; water has highest specific heat of any known substance.

Specific Pesticide see SELECTIVE PESTICIDE.

Spectrum A visual display, photographic record, or plot of the distribution of the intensity of a given type of radiation as a function of its wavelength, energy, frequency, momentum, mass, or any related quantity.

Spent (Depleted) Fuel Nuclear reactor fuel that has been used to the extent that it can no longer effectively sustain a chain reaction.

Spill The accidental release of contaminating materials.

Spillage Any escape, leakage, dripping, or running over of materials.

Spiller see RESPONSIBLE PARTY.

Splash Guard Deflecting shield sometimes installed on tank trailers to protect meters, valves, etc.

Splice Bar see JOINT BAR.

Split Sample Subsamples that do not differ significantly from each other or from the original sample; used to compare performance between/among laboratories.

Splitter Valve Valve installation to divide pipeline manifold.

Spontaneous Fission Fission occurring without an external stimulus; process occurs occasionally in all fissionable materials.

Spontaneous Ignition A material proceeding without constraint by internal impulse or outside energy to kindle or set fire; quick or slow oxidation or combustion brought about by chemical, elec-

186

trical, biological (bacterial), or physical processes (vibration, pressure, friction) without assistance of extraneous sources of heat (flame, sparks, hot or glowing bodies).

Spontaneously Combustible Material able to increase in temperature to a point of ignition without drawing heat from its surroundings.

Spot Place a RR car in a designated position or specific location, such as at a warehouse door, usually for loading or unloading.

Spot System A classification system for repair work where RR cars are moved progressively from one spot to another with different items of work done by a unit force.

Spot Treatment Pesticide application directed at small area such as at specific plants; opposite of general treatment.

Spray Drift see DRIFT.

Spread Tandem A two-axle assembly wherein the axles are spaced to allow maximum axle loads under existing regulations; distance between center of axles of a spread tandem generally has been more than 50 inches.

Spreader An adjuvant that increases the area that a given volume of liquid will cover on a solid surface.

Spreader Bars Transverse rigid stabilizing rods attached to the side walls of a trailer; common usage is on open top van trailers.

Spring Deflection Depression of a trailer suspension spring when the spring is placed under load.

Spring Seat Suspension component used to support and locate spring on axle; formerly called a spring chair.

Spring Switch RR switch equipped with a spring mechanism arranged to restore the switch points to normal position after having been trailed through.

Spur Track A stub RR track extending out from a main track.

Stabilization (1) Stage of an incident when the immediate problem or emergency has been controlled, contained, or extinguished. (2) Hazardous waste chemical treatment method by which a chemical reaction produces an insoluble form of the waste or incorporates the waste into a form that is insoluble.

Stabilization Lagoon A shallow pond for the storage of waste water before discharge to a stream or to a treatment facility.

Stabilizer Inhibitor.

187

Stabilizing Jack A device usually of single leg construction used on the forward end of a trailer as an auxiliary support to prevent nosediving.

Stable (1) Incapable of spontaneous change. (2) Not radioactive.

Stage of the Incident One of five definite, identifiable phases through which an emergency passes from onset (interruption of normal conditions) to stabilization.

Staging Area (ICS) Location where incident personnel and equipment are assigned in anticipation of immediate assignment to combat operations.

Standard Gauge Distance of four feet, eight and one-half inches between the heads of the rails.

Standard Operating Procedures Detailed instructions for implementation of emergency plans by the various response agencies.

Standard Rate Rate established via direct routes from one point to another.

Standard Route (RR) The line or lines that maintain standard rates.

Standard Transportation Commodity Code (STCC Code Number) A listing of code numbers for categories of articles being shipped, in general use by carriers.

State Agency Coordinator (SAC) The representative responsible for coordinating resources and maintaining liaison with the federal on-scene-coordinator. Obtains and provides pertinent information for all state agencies and provides assistance to the incident commander.

State Board California Water Resources Control Board.

State Law Abstracts of state laws, regulations, and standards concerning hazardous substances.

State Override Refers to procedures whereby local decisions on hazardous waste management facility siting can be overturned by a state agency.

State Preemption Refers to the preemption of local decision-making authority over hazardous waste management facility siting by state agency so that no local decision is required to site such facilities.

State Siting Regulations Agency-issued, state-authorized directives to implement RCRA dealing with the approval or restriction of facility locations and/or permitting processes.

State Warning Center The center within the state's Office of Emergency Services that monitors seismic activities and is the reporting office for any release or threatened release of a hazardous material or spill.

State Water Resources Control Board (SWRCB)   The State Board, established by the Porter-Cologne Water Quality Control Act of 1967; adopted the California state policy for water quality control in accordance with legislative policies; designated as the state water pollution control agency for the purpose of the federal Water Pollution Control Act; reviews actions of the Regional Water Quality Control Boards relative to regulation of waste discharges.

Stationary Installation (Fixed or Permanent Installation)   An installation of LP-Gas containers, piping, and equipment for use indefinitely at a particular location; an installation not normally expected to change in status, condition, or place.

Stationary Source   A fixed facility from which a release of hazardous material may originate.

Statistically Significant   When the difference between a predicted and an observed value is so large that it is improbable that it could be attributed to chance.

Status Under Law   Lists all federal regulations, rules, standards, NIOSH recommendations, and agency lists for scientific evaluation and potential regulation.

Stay Put   A resident in a hazardous or potentially hazardous area who refuses to relocate during an incident; a person who is too infirm to be moved and must be protected in place.

Stay Time   Period of time personnel may work within a radiation area before receiving a dose equal to the field administrative limit; twenty-five rems recommended.

Steel   Any of various hard, strong, durable, malleable alloys of iron and carbon, usually containing between 0.2 and 0.7% carbon, often with other constituents such as manganese, nickel, copper, tungsten, cobalt, or silicon and widely used as a structural material.

Step-Off Area   Area designated for entry to and exit from a radiation area.

Sterilization   The use of physical or chemical procedures to destroy all microbial life, including highly resistant bacterial endospores; more thorough than disinfection.

Sticker   An adjuvant that increases the adherence of a pesticide.

Stock Car   A car for the transportation of livestock equipped with slatted sides, single or double deck, and sometimes feed and water troughs.

189

Stolon   An above-ground stem that produces roots.

Stomach Poison   A pesticide that kills an animal when eaten or swallowed.

Stopping in Transit   The holding of a shipment by a carrier on order of the owner after the transportation movement has started and before it is completed.

Stopping Power   A measure of the effect of a substance upon the kinetic energy of a charged particle passing through it; see ABSORPTION.

Storage   Containment of waste on a temporary basis in such a manner as not to constitute disposal of such waste; containment of such waste for over a year constitutes disposal.

Storage Facility   Any location storing wastes for less than ninety days for subsequent transport off-site (except generators who store their own wastes).

Storage Tank   Any manufactured nonportable covered device used for containing pumpable hazardous wastes.

Storage in Transit   The stopping of freight at a point located between the point of origin and destination to be stored and forwarded at a later date.

Stratigraphy   The arrangement of soil into layers or strata.

Streaming   The channeling of radiation into a beam; occurs if the lid of a shielding cask is removed, allowing radiation to escape the container; the radiation would be in the form of a column, similar to the beam of a flashlight.

Stress   (1) (Verb) To apply a form of force or energy that tends to strain, deform, or otherwise change a body or mass. (2) (Noun) The condition of being subjected to force or energy tending to deform or strain. (3) A state of tension put on or in a shipping container by internal chemical action, external mechanical damage, or external flames or heat.

Stressee   The actor whose state is being changed by stresses.

Stressor   The actor applying the force or energy being transferred.

Strict Liability   The responsible party is liable even though they have exercised reasonable care.

String   Two or more RR freight cars coupled together.

Structures   The basic installations and facilities on which a community depends (roads, powerplants, utilities, communication systems).

Stub Track   A track connected at one end only.

Subacute Toxicity   How poisonous a pesticide is to an animal or person after repeated doses (long-term exposure).

Subcritical Mass   An amount of fissionable material insufficient in quantity or of improper geometry to sustain a fission chain reaction.

Subframe   Assembly employed to unit trailer body and suspension.

Sublimination   Changing of a substance from a solid to a gaseous state without ever going through the liquid phase.

Submersion Skimmer   In oil operations, a type of mechanical skimmer incorporating a moving belt inclined at an angle to the water surface in such a way that oil in the path of the device is forced beneath the surface and subsequently rises (due to its buoyancy) into a collection well.

Subsurface Contamination   Any type of contamination located below the ground surface.

Substrate   Materials that form the base of something. In biology, the base on which an organism lives. Substrate materials include water, soils, and rocks as well as other plants and animals.

Suction Lysimeter   A device for extracting liquid samples from the unsaturated zone.

Suction Skimmer   A type of mechanical skimmer incorporating an enlarged intake device at the end of a vacuum hose to increase the surface area over which suction of an external pump is exerted.

Suggested Control Measures   Procedures necessary to attain federal and state ambient air quality standards.

Suit in Equity   To ask the courts to order someone to do something other than pay money. DER has the authority to ask a court to order a hazardous waste handler to do something it is supposed to do, or to not do something it is not supposed to do (e.g., pollute).

Sump   (1) The low point of a tank at which the emergency valve or outlet valve is attached. (2) A stationary device designed to contain an accumulation of hazardous waste resulting from a hazardous discharge from a tank, container, waste pile, surface impoundment, landfill, or other hazardous waste management structure.

Superfund   The Comprehensive Environmental Response, Compensation and Liability Act of 1980 provides the federal government with the mechanism to take emergency or remedial action to clean up both abandoned and existing disposal sites whenever there is a release or potential release of a hazardous substance which may present imminent and substantial danger to public health and welfare; funds for these cleanup actions come from a $1.6 billion trust fund called the "Response Trust Fund."

Superheating   Heating of a vapor, particularly saturated steam, to a temperature much higher than the boiling point at the existing pressure; occurs in power plants to improve efficiency and to reduce condensation in the turbines.

Superior Train   A train having precedence over another train.

Supplement   Any substance added to a pesticide to improve its performance; adjuvant.

Supplies   Equipment and all expendable items assigned to an incident.

Supply Officer   The individual in a command post responsible for movement, requisitioning, and issuing of materials, clothing, or equipment to be used at the scene of an emergency incident; also called supply unit leader.

Support Zone   The area outside of the warm zone; equipment and personnel are not expected to become contaminated in this area. Area where resources are assembled to support the hazardous materials operation.

Supportive Care (EMS)   Measures taken to maintain vital body functions.

Supports   Devices generally adjustable in height, used to support the front end of a semitrailer in an approximately level position when disconnected from towing vehicle.

Surety Bond   A guarantee of performance to complete a contract or obligation.

Surface Active Agents   Chemicals that alter the forces of surface tension between adjacent molecules; generally decrease the surface tension of a fluid such as an oil, used to facilitate its dispersion throughout the water column.

Surface Impoundment   Facility or part of a facility that is a natural topographic depression, man-made excavation, or diked area formed primarily of earthen materials (although may be lined with synthetic materials) designed to hold an accumulation of liquid wastes or wastes containing free liquids, and which is not an injection well; examples: holding, storage, settling, and aeration pits, ponds, and lagoons.

Surface Resistivity (Electric Soil Survey)   A technique that measures relative values of the earth's electrical resistivity; used to define subsurface geologic and hydrologic conditions.

192

Surface Seal   A shortened annular seal used to ensure no contaminants will enter the annular space from the ground surface.

Surface Spray   A pesticide sprayed evenly over the entire outside of the object to be protected.

Surface Tension   Force of attraction among the surface molecules of liquid; affects the rate at which a spilled liquid will spread over a land or water surface into the ground.

Surface Water   Water located above ground, such as ponds, lakes, streams, rivers, etc.

Surfactant   An adjuvant that improves the emulsifying, dispersing, spreading, and wetting properties of a pesticide.

Surveillance   Periodic medical examinations required by OSHA standards; listed by individual medical tests and procedures.

Survey Meter   An instrument designed to detect and measure radiation.

Susceptibility   Degree to which an organism can be injured or affected by a pesticide at a known dosage (or exposure).

Susceptible   Capable of being injured, diseased, or poisoned.

Susceptible Species   A plant or animal that is poisoned by moderate amounts of a pesticide.

Suspended   (1) Dispersed in a liquid. (2) A pesticide use no longer legal, whose remaining stocks cannot be used (more severe than Cancelled).

Suspension   (1) An assembly employed to connect axle to subframe in such a manner as to cushion road shock and to carry the load. (2) Combination of a substance with a liquid in which the substance is not dissolved. (3) Pesticide formulation in which finely divided solid particles of an active ingredient are mixed in a liquid.

Swab   Device that fits the inside of tubing closely and is pulled through to clean the tubing, or to pull such a device through the tubing.

Swale   A slight longitudinal depression in the midst of generally level land.

Swamper   Helper on a truck.

Swatch   The width of ground covered by a sprayer when it moves across a field or other treated area.

Sweet (Gas)   Gas that has the hydrogen sulphide removed; *see* SOUR GAS.

193

Switch (1) Connection between two lines of RR track to permit cars or trains to pass from one track to the other. (2) To move RR cars from one place to another within a yard, industry, or terminal.

Switch Back A series of zigzag curves in mountainous terrain designed to reduce the rate of climb or descent.

Switch Engine Locomotive used for moving cars in terminals and yards; usually built to carry all its weight on the driving wheels.

Switch List (RR) A list of freight cars in track standing order showing cars by initial, number, type of car, and showing where cars are to be switched as required by local practice.

Switch Lock A fastener, usually a spring padlock, used to secure the switch or derail stand in place.

Switch Order Instructions to move a RR car from one place to another within switching limits.

Switch Stand Device by which a switch is thrown, locked, and its position indicated; consists of base, spindle, lever, and connecting rod and usually has a lamp and a banner signal.

Switch Target A visual day signal fixed on the spindle of a switch stand, or the circular flaring collar fitted around the switch lamp lens and painted a distinctive color to indicate the position of the switch.

Switch Tender Employee responsible for aligning tracks for engine and car movements by throwing switches.

SWRCB (or State Board) State Water Resources Control Board.

Symptom A warning that something is wrong; a feeling of being sick; an indication of poisoning or disease in a person, animal, or plant.

Syndrome A series of symptoms associated with a specific illness.

Synergistic When the combined action of two or more pesticides is greater than the sum of their activity when used alone.

Synonym Common name, slang name, acronym, trade name, brand name, or alternate chemical name.

Synthetic Organic Pesticides Man-made pesticides that contain carbon, hydrogen, and other elements.

Synthetic Organic Sorbent One of several organic polymers, genally in the form of plastic foams or plastic fibers, used to recover spilled oil; have higher recovery capacities than either natural organic sorbents or mineral-based sorbents and many can be reused after oil is squeezed out. Examples: polyurethane foam, polyethylene, and polypropylene.

System Organization and Management Element (EMS) The structure of an EMS system; includes administration, policy making, planning, data collection and analysis, evaluation, and financing.

Systemic A pesticide absorbed by one part of a plant or animal and moved to another section where it acts against a pest; example: A systemic insecticide can be applied to the soil, be absorbed by the plant's roots, move into the leaves, and then control insects when they feed on the leaves.

Systemic Insecticide Pesticide absorbed into the plant or animal to be protected, in order to control the attacking pest.

Systemic Toxic Exposure Toxin affects the body as a whole; spread through the bloodstream.

T

Tactics Successful methods or procedures used to deploy various resources to achieve objectives.

Take Internally Ingest; to eat or swallow; to take by mouth into the digestive system.

Tamper A power-driven machine for compacting ballast.

Tandem Two-axle suspension.

Tank A large receptacle for holding, transporting, or storing liquids. A stationary device designed to contain an accumulation of hazardous waste, constructed primarily or entirely of nonearthen materials such as concrete, steel, or plastic which provides structural support and containment.

Tank Backfill Material used to backfill the excavation surrounding an underground storage tank.

Tank Car A RR car used for carrying liquids, such as oil, molasses, vinegar, acid, etc.

Tank Dome A vertical cylinder attached to the top of a tank car, permitting tank proper to be filled to full cubical capacity, allowing room for expansion of the contents into the dome.

Tap Line (RR slang) A short railroad usually owned or controlled by the industries which it serves and connecting with a trunk line.

Tar   A black or brown hydrocarbon material ranging in consistency from a heavy liquid to a solid; most common source of tar is the residue left after fractional distillation of crude oil.

Tar Balls   Compact semi-solid or solid masses of highly weathered oil formed through the aggregation of viscous, high carbon-number hydrocarbons with debris present in the water column; generally sink to the sea bottom but may be deposited on shorelines where they tend to resist further weathering.

Tare Weight   (1) Weight of a container and the material used for packing. (2) Weight of any empty freight car.

Target   The area, building, animal, plant, or pest that is to be treated with a pesticide.

Target Pest   The pest at which a pesticide application or other control method is directed.

Task Force (ICS)   A group of suppression and rescue resources temporarily assembled for a specific mission.

Team Leader   *see* ENTRY TEAM LEADER.

Team Track   A RR track on which cars are placed for public use to load or unload freight.

Technical Assistance   Personnel, agencies, or printed materials providing technical information on the handling of hazardous materials.

Technical Assistance Grants   Funds made available to communities to hire technical experts to assist in the review of hazardous waste management facility proposals.

Technical Material   In pesticides, the active ingredient as it is manufactured by a chemical company before formulation.

Technical Pesticide   Highly concentrated pesticide which is to be combined with other materials to formulate pesticidal products.

Technical Specialist   Individual or unit responsible for the collection, evaluation, and dissemination of information concerning specialized data.

Telecommunications   Pertaining to the transmission of signals over long distances, such as by telegraph, radio, or television.

Telemetry   (1) System by which the variations recorded by any physical or other instrument can be shown at a distance by means of electricity. (2) (EMS) Transmission of digital and analog biomedical data.

Temperature   A condition of hotness or coldness whose state is being described and is relative only to the condition being observed. Measured artificially by the scales of different types, i.e., Fahrenheit, Centigrade, Kelvin, Rankin, etc. Absence of heat is manifested as cold temperatures; presence of heat is referred to as high temperatures.

Tension Member The part of a floating containment boom that carries the load placed on the barrier by wind, wave, and current forces; commonly constructed from wire cable due to its strength and stretch resistance.

Teratogen A substance that may cause abnormal development in unborn animals.

Teratogenesis Alteration in the formation of cells, tissues, and organs resulting from physiologic and biochemical changes on fetus during growth; may affect function as well as structure of developing cells (occurs very early in fetal period).

Terminal Carrier The railroad making delivery of a shipment at its destination.

Terminal Charge (RR) A charge made for services performed at terminals.

Terminals (1) Points where employees in road train and engine service originate and/or terminate their tour of duty. (2) A device attached to the end of a wire or cable or to an electrical apparatus for convenience in making connections.

Termination That portion of incident management in which personnel are involved in documenting safety procedures, site operations, hazards faced, and lessons learned from the incident. Termination is divided into three phases: debriefing, post-incident analysis and critique (NFPA 472, 1–3). *see* POST-INCIDENT ANALYSIS.

Test Animals Laboratory animals exposed to pesticides so that toxicity and hazards can be determined.

Test Weight Car (RR) *see* SCALE TEST CAR.

Tests Medical diagnoses used as an index of exposure to a specific substance.

Tetra- A prefix indicating the number four.

Therapeutic Care (EMS) Measures taken to relieve or cure bodily dysfunctions resulting from injury or illness.

Thermal Of, about, or related to heat.

Thermal Treatment Process by which hazardous waste is rendered nonhazardous or is reduced in volume by exposing the waste to high temperatures; organic materials are oxidized and converted to carbon dioxide and water.

Thermoluminescent Dosimeter (TLD)   A device used for measuring beta, gamma, and X-ray radiation exposure; energy absorbed from the radiation raises the molecules of the material in the TLD to excited states where it remains until it is heated; at that time it gives off an amount of light proportional to the dose, and that light is measured with a photomultiplier tube.

Thermometer   A device for measuring heat intensity.

Thermonuclear Reaction   A process by which very high temperatures bring about the fusion of two light nuclei to form the nucleus of a heavier atom, releasing a large amount of energy; in a hydrogen bomb, the high temperature to initiate the thermonuclear reaction is produced by a preliminary fission reaction.

Thieving Rod   A glass rod used like a coliwassa, except the liquid is contained in the tube by a vacuum pressure.

Third Rail (RR)   An electric conductor located alongside the running rail from which power is collected by means of a sliding contact shoe attached to the truck of electric equipment.

Third Rail Shoe (RR electric locomotive)   An insulated metallic sliding contact, mounted on the truck of an electric locomotive for collecting current from an insulated third rail located alongside the running rails. Positive contact between shoe and rail is maintained by gravity, a spring, or pneumatic pressure.

Thorium   A naturally radioactive element with atomic number 90 and, as found in nature, an atomic weight of approximately 232. The fertile thorium 232 isotope is abundant and can be transmitted to fissionable uranium 233 by neutron irradiation.

Threshold   The point at which a physiological or toxicological effect begins to be produced by the smallest degree of stimulation.

Threshold Dose   The minimum exposure dose of a chemical that will evoke a stated or nontoxicological response.

Threshold Limit Value   The values for airborne toxic materials that are to be used as guides in the control of health hazards and represent concentrations to which nearly all workers may be exposed eight hours per day over extended periods of time without adverse effects.

Threshold Limit Value–Ceiling (TLV– C)   The concentration that should not be exceeded during any part of the working exposure.

Threshold Limit Value–Time Weighted Average (TLV–TWA)   An exposure level under which most people can work consistently for eight hours a day, day after day, with no harmful effects.

Title III   The Emergency Planning and Community Right-to-Know Act of 1986; specifies requirements for organizing the planning process at the state and local levels for specified extremely hazardous substances; minimum plan content: requirements for fixed facility owners and operators to inform officials about extremely hazardous substances present at the facilities, and mechanisms for making information about extremely hazardous substances available to citizens.

TLV   see THRESHOLD LIMIT VALUE.

TMP   Tank monitoring program.

Totally Encapsulated Suits   Special protective suits made of material that prevents toxic or corrosive substances or vapors from coming in contact with the body; see FULLY ENCAPSULATED SUIT.

Toxic   Poisonous; relating to or caused by toxin; able to cause injury by contact or systemic action to plants, animals, or people.

Toxic Air Contaminant   An air pollutant that may cause or contribute to an increase in mortality or an increase in serious illness, or that may pose a present or potential hazard to human health (Section 39655, Health and Safety Code).

Toxicity   The degree of being poisonous; capability of a poisonous compound to produce deleterious effects in organisms such as alteration to behavioral patterns or biological productivity or death.

Toxic Chemicals   EPA uses this term for chemicals whose total emissions and releases must be reported annually by owners and operators of certain facilities that manufacture, process or otherwise use a listed toxic chemical as identified in SARA Title III.

Toxic Waste   Refuse posing a substantial present or potential hazard to human health or to the environment when improperly managed; includes poisonous wastes, carcinogenic, mutagenic, teratogenic, or phytotoxic wastes, or wastes toxic to aquatic species.

Toxin   A poison produced by a plant or animal.

TPH   Total Petroleum Hydrocarbons.

Tracer   (1) To discover by going backwards over the trail of (something). (2) With freight, to search for a shipment to expedite movement or establish delivery.

Track (RR)   The space between the rails and space of not less than four feet outside each rail.

Track Car (RR)   A self-propelled car which may not operate signals or shunt track circuits; includes burro cranes, detector cars, weed burners, tie tampers, etc.

Track Circuit An electrical circuit including the rails and wheels of the train; used for controlling signal devices (fixed signals as well as flashers and gates at crossings).

Trade Name Brand name.

Traffic Control System A block signal system under which train movements are authorized by block signals, whose indications supersede the superiority of trains for both opposing and following movements on the same track.

Traffic Control Point (TCP) Places along a movement route that are manned by emergency personnel to direct and control the flow of traffic.

Trailing Movement Movement of a train over points of a switch which face in the direction in which the train is moving.

Trailing Point Switch A switch on which the points face away from approaching traffic.

Train An engine or more than one engine coupled, with or without cars, displaying a marker.

Train Line The complete line of air brake pipes in a train, including the rigid piping secured under cars and flexible connections between cars and the locomotive.

Train of Superior Direction Train given precedence in the direction specified by timetable as between opposing trains same class.

Train of Superior Right Train given precedence by timetable.

Transfer Agreement (EMS) Written contract between health facilities providing reasonable assurance concerning transfer of patients between facilities when medically appropriate as determined by the attending physician.

Transfer Station A facility designed to accept wastes from the surrounding region for purpose of temporary storage and/or repackaging; when wastes have been collected in sufficient quantity to make shipment economically practical, they are transported in larger trucks in a compact and orderly manner to disposal site or resource/recovery facility.

Transformation Changing from one form to another; transmutation.

Transient Rapid change of a plant operating parameter such as temperature, pressure, or steam generator water level.

Transition Section of a tank joining two unequal cross sections.

Transit Privilege   Service granted to a shipment en route such as milling, compressing, refining, etc.

Transit Rate   A rate restricted in its application to traffic which has been or will be milled, stored, or otherwise specially treated in transit.

Translocated Pesticide   A pesticide that moves within a plant or animal after it has entered by some path; a systemic pesticide.

Transmissivity   The transmission rate of water (based on a unit width of an aquifer) relative to a hydraulic gradient.

Transmitter   Apparatus for production and modulation of radio frequency energy for purpose of radio communication.

Transmutation   The changing of one element into another by a nuclear reaction or series of reactions; example: the transmutation of uranium-238 into plutonium-239 by absorption of a neutron.

Transport   To carry from one location to another.

Transport Mode   Method of transportation (highway, rail, water, pipelines, air).

Transport Time (EMS)   The interval of time required for emergency medical transport of patient from the scene of an incident to a receiving facility.

Transportation Component (EMS)   Combined personnel, equipment resources, and arrangements to deliver initial emergency medical care to the location of a patient, rescue the patient from a hazardous environmental situation, and then transport that patient to the appropriate medical facility.

Transposing   The act of rearranging numbers or letters from their proper order.

Treatment   Any method, technique, or process that changes the physical, chemical, or biological character or composition of any hazardous waste or removes or reduces its harmful properties or characteristics for any purpose.

Treatment Facility   Any facility at which hazardous waste is subjected to treatment or where a resource is recovered from a hazardous waste (Section 66220, Title 22, California Administrative Code).

Tremie Pipe   A pipe used to fill the annular space (space between soil and outside of well casing) from the bottom up when completing a well installation or when sealing an abandoned well.

Tri-   A prefix that indicates number three.

Triage   The determination of priorities in an emergency.

Trip or Travel Blank   A sample container filled in the laboratory with organic free water and carried unopened during the sampling trip. It must be prepared by the laboratory supplying sample containers. It is used to identify contamination introduced from the originating laboratory. The trip blank remains with the collected samples and is analyzed along with the field samples to check residual contamination. Trip blanks are mandatory for volatile hydrocarbon analysis in water.

TRPH   Total Recoverable Petroleum Hydrocarbons.

U

UEL   Upper explosive limit.

μg   Microgram; 1/1000 of a milligram.

μg/g   Microgram/gram.

μg/l   Microgram/liter.

μg/ml   Microgram/milliliter.

UHF (Ultra-High Frequency)   Radio spectrum between 300 and 3000 MHz; includes EMS channels from 450 to 470 MHz.

μl   Microliter; 1/1,000,000 of a liter.

Ultra-Low Volume (ULV)   Spray application of a pesticide that is almost pure active ingredient, sprayed over a large area.

Ultraviolet Radiation   The portion of the electromagnetic spectrum emitted by the sun adjacent to the violet end of the visible light range; often called "black light," it is invisible to the human eye but when it falls on certain surfaces it causes them to fluoresce or emit visible light; responsible for the photo-oxidation of certain compounds including hydrocarbons.

Unassigned Car   RR car, usually with some interior loading devices, but not assigned to a particular industry or commodity.

Unauthorized Release   Any unpermitted release or leak of hazardous materials into the environment.

Uncertainty Factor (UF)   An approach to the determination of exposure levels based on the application of uncertainty factors to the maximum dose level causing no observable adverse effects.

Unclaimed Freight   Freight shipment that has not been called for by consignee or owner.

Unconfined Aquifer   An aquifer whose upper level can extend to ground surface.

Uncoupling Lever, Uncoupling Rod (RR)   A device to uncouple RR cars without moving between them, this rod has a bent handle forming a lever and is usually attached to the end sill; lever proper is attached to the rod and operates the unlocking mechanism, but in the case of freight cars the lever and rod are generally constructed in one piece.

Undercarriage   Complete subframe, suspension with one or more axles which may be interconnected, and wheels, tires, and brakes.

Underground Storage Tank (UST)   Any hazardous material containment device of nonearthen material partially or substantially located below ground, including associated piping.

Underground Water   Water and waterways located below soil surface.

UNI   Ente Nazionale Italiano di Unifacazione (Italy).

Uniform Coverage   Even application of pesticide over an entire area, plant, or animal.

Uniform Demurrage Rules   Schedules providing rules and charges for demurrage which are, in general, used throughout the United States, having the approval of but not prescribed by the ICC.

Uniform Freight Classification   A listing of commodities showing their assigned class rating to be used in determining freight rates, together with governing rules and regulations.

Unique Site Feature   Physical characteristics of the site that could influence the movement and direction of contaminants through the subsurface.

Unit (ICS)   Organizational element having functional responsibility for a specific incident planning, logistic, or finance activity.

UN   United Nations.

United Nations Identification Number (UN)   When UN precedes a four-digit number, it indicates this identification number is used internationally to identify a hazardous material.

Universal Cylinder   A DOT cylinder specification container, constructed and fitted with appurtenances in such a manner that it may be connected for service with its longitudinal axis in either vertical or horizontal position, and so that its fixed maximum liquid level gauge, relief device(s), and withdrawal appurtenance will function properly in either position.

Universal Precautions   A system of infectious disease control that assumes that every direct contact with body fluids is infectious and requires every employee exposed to direct contact with body fluids to be protected as though such body fluids were HBV- or HIV-infected.

Unsaturated Zone   Region between the land surface and the upper boundary of the zone of saturation or water table; zone of aeration.

Unstable   Ready to behave erratically or in an unwanted manner.

Unstable Materials   Substances capable of rapidly undergoing chemical changes or decomposition.

Unusual Event   Off-normal events that do not by themselves constitute significant events at a nuclear facility, but could indicate a potential degradation in the level of safety.

Upper Coupler Assembly   An upper coupler plate, reinforcement framing, and fifth wheel kingpin mounted on a semitrailer; formerly called upper fifth wheel assembly.

Upper Explosive Limit (UEL)   The highest concentration of the material in air that can be detonated.

Upwind   In or toward the direction from which the wind blows.

Uranium   The basic raw material of nuclear energy, uranium is a radioactive element with the atomic number 92 and, as found in natural ores, an average atomic weight of approximately 238. The two principal natural isotopes are uranium 235 (0.7% of natural uranium), which is fissionable, and uranium 238 (99.3% of natural uranium), which is fertile. Natural uranium also includes a minute amount of uranium 234.

Uranium Hexafluoride   A volatile compound of uranium and fluorine; the process fluid in the gaseous diffusion process.

Uranium Tetrafluoride   A solid green compound called green salt; an intermediate product in the production of uranium hexafluoride.

Uranium Trioxide   Orange Oxide; an intermediate product in the refining of uranium.

Used Motor Oil   Any oil previously used in any machinery; main markets are in road oiling, industrial fuel, and rerefining.

USDA   United States Department of Agriculture.

UV   Ultraviolet (light wavelength).

V

Vadose Zone The unsaturated area between ground surface and the water table.

Vapor Gases given off, with or without the aid of heat, by substances that under ordinary circumstances are either a solid or a liquid; gas, stream, mist, fog, fume, or smoke.

Vapor Density Measurement of the weight of vapor as compared with an equal volume of dry air; a figure of less than one indicates a vapor lighter than air which will rise; vapor density greater than one tends to settle; an important consideration when dealing with potential or actual fires where flammable liquids and gases are involved, as most flammable liquid vapors have vapor densities of greater than one, such as gasoline at 1.4 and ethyl alcohol at 1.9, while some flammable gases have vapor densities lighter than air, such as hydrogen at 0.6 and acetylene at 0.9.

Vapor Dispersion The movement of vapor clouds in air due to turbulence, gravity spreading, and mixing.

Vapor Drift *see* DRIFT.

Vapor Pressure The property that determines how easily something evaporates; the lower the vapor pressure, the more easily a substance will evaporate. Also, the pressure exerted by the molecules of a liquid leaving the liquid and entering the void space of a closed container; measured in terms of psi. Increased temperatures applied to a closed container will cause the vapor pressure to increase; when the vapor pressure within a closed container exceeds the capability of the container to withstand internal pressure, a mechanical rupture of the container will occur; *see* BLEVE.

Vapor Protective Suit *see* LEVELS OF PROTECTION.

Vaporization The process of becoming a gas.

Vaporize To evaporate; to form a gas and disappear into the air.

Vaporizer A device for converting liquid to vapor by means other than atmospheric heat transfer.

Vaporizer, Direct Fired In LP-Gas, a vaporizer in which heat furnished by a flame is directly applied to some form of heat exchange surface in contact with the liquid LP-Gas to be vaporized.

Vaporizer, Indirect A vaporizer in which heat furnished by steam, hot water, or other heating medium is applied to a vaporizing chamber or to tubing pipe, coils, or other heat exchange surface containing the liquid LP-Gas to be vaporized; the heating of the medium used being at a point remote from the vaporizer.

VOA Volatile organic analysis.

VOC Volatile organic compound.

Vulnerability The susceptibility of life, the environment, and/or property to damage by a hazard.

Waiting Period Time interval.

Warm Zone The area where personnel and equipment decontamination and hot zone support takes place. It includes control points for the access corridor and thus assists in reducing the spread of contamination. Also referred to in other documents as the decontamination zone, contamination reduction, yellow zone, support zone or limited access zone (NFPA 472, 1–3).

Warning Signal word used on pesticide labels to identify moderately toxic substance.

Wash To clean, usually with clear water.

Waste Useless or discarded materials.

Waste Discharge Requirements An authorization by the Regional Water Quality Control Boards setting requirements for discharge of wastes from point sources.

Waste Exchange Waste clearinghouses where pretreated or untreated hazardous wastes are transferred; operating on the principle that "one man's waste can be another's feedstock."

Waste Management The total process of waste collection, from its point of generation through its transportation, treatment, and final acceptable disposal.

Waste Recycling Processes and procedures that allow industries to recover chemical resources for new or additional uses.

Waste Reduction Modifications in industrial processes that result in the production of fewer hazardous by-products; prevention of waste at its sources, either by redesigning of products or by changing societal patterns of production and consumption.

Waste Transfer Center A reception area used as an adjunct to a waste collection system; may be fixed or mobile.

Water Reactive   Having properties of reacting violently when contacted by water, generating extreme heat, burning, exploding, or rapidly reacting to produce an ignitable, toxic, or corrosive mist, vapor, or gas.

Water Table   The top of the saturated zone where unconfined ground water is under atmospheric pressure.

Weight Filling   Filling containers by weighing the LP-Gas in the container. No temperature determination or correction is required, since a unit of weight is a constant quantity regardless of temperature.

Weir Skimmer   A type of skimmer that employs the force of gravity to drain oil from the water surface; basic components include a weir or dam, a holding tank, and an external pump. As oil on the water surface falls over the weir or is forced over by currents into the holding tank, it is continuously removed by the pump.

Well   A driven, drilled, bored, or dug excavation, lower inclined from the vertical, with a depth greater than the largest surface dimension, generally of cylindrical form, and often walled by some means to prevent the excavation from caving in.

Well Car   *see* DEPRESSED CENTER FLAT CAR.

Well Centralizer   A device used to center a well casing in a borehole.

Well Development   The procedure used to assure stabilization of the gravel pack envelope and adjacent formation following well construction.

Well Log   A record of installation of a well. It includes construction specifications of the well, depth, owner, location, and a description of the soil profile; prepared by the well driller. Well log records are maintained by the State Department of Water Resources, some county agencies, and the U.S. Geological Survey.

Well Purging   The procedure used to assure removal of standing water prior to sampling of representative, formational groundwater.

Wetlands   Areas where water is at, near, or above the land surface long enough to be capable of supporting aquatic or hydrophytic vegetation and have soils indicative of wet conditions.

Wettable Powder   Finely ground pesticide dust that will mix with water for form a suspension for application; may not burn, but may release toxic fumes under fire conditions.

Wetting Agent   An adjuvant that reduces surface tension and allows a pesticide to spread out and more evenly coat a surface; it decreases the rolling and running off of a pesticide.

Wheel Flange   The projecting edge or rim on the circumference of a RR car wheel for keeping it on the rail.

Whole Body Counter   Device used to identify and measure the radiation in the body of people and animals; uses heavy shielding to keep out background radiation and ultrasensitive scintillation detectors and electronic equipment.

Whole Body Exposure   The exposure of the entire body to radiation, rather than an isolated part such as an arm, foot or head.

Wicking Agent   Substances such as straw, wood chips, glass beads, and treated silica that are used to increase oxygen availability and provide insulation between oil and water during the disposal of spilled oil by burning.

WIP   Well Investigation Program; conducted by the Regional Water Quality Control Board.

Withdrawal   Water pumped out of a well.

Wood Preservatives   Pesticides used to treat any wood to prevent insect damage, dry rot, or other damage.

Wind Sock   Cone-shaped cloth sock which hands in an open area of the airport to serve as a weather vane, indicating wind direction.

Wireline   Wire rope.

Wire Rope   Rope used in the oil industry composed of steel wires twisted into strands that are in turn twisted around a central core of hemp or other fiber to create a rope of great strength and considerable flexibility.

Wood Preservatives   Pesticides used to treat any wood to prevent insect damage, dry rot, or other damage.

Wood Pulp   Primary material from which most papers are made; made of small, loose wood fibers mixed with water.

Working Pressure   Maximum pressure at which an item is to be used at a specified temperature.

X

X   The unnamed SI unit for measuring radiation exposure; one unit = one coulomb/kg = 3876 roentgens.

X-Ray   A penetrating form of electromagnetic radiation emitted either when the inner orbital electrons of an excited atom return to their normal state (these are characteristic X-rays), or when a metal target is bombarded with high speed electrons (bremsstrahlung). X-rays are always nonnuclear in origin.

Y

Yard (RR) A system of tracks branching from a common lead or ladder track, with defined limits, used for switching, making up trains, or storing of cars.

Yard Clerk (RR) Person engaged in clerical work in and around RR yards and terminal.

Yarding in Transit Unloading, storing, sorting, etc., of forest products in transit.

Yield The total energy released in a nuclear explosion; usually expressed in equivalent tons of TNT (the quantity of TNT required to produce a corresponding amount of energy).

-yl A suffix used in forming the names of radicals, usually univalent radicals, such as alkyl or acyl.

-ylene A suffix used in forming the names of unsaturated hydrocarbons, such as ethylene; may be used in forming the names of bivalent radicals with the free valences on different carbon atoms, such as phenylene.

-yne A suffix indicating unsaturation of one acetylenic such as pentyne. Modern nomenclature prefers this suffic to that of -ine.

Z

Z The symbol for Atomic Number.

Zero Tolerance (Obsolete) Where, by law, no detectable amount of pesticide may remain on any agricultural commodities when offered for shipment. Extremely sensitive detection methods now available have forced change in law.

Zone As applied to oil reservoirs, describes an interval which has one or more distinguishing characteristics, such as lithology, porosity, saturation, etc.

Zone of Influence Maximum extent to which a waste disposal facility will affect surface and groundwater quality.